# Practical Cattle Farming

# Practical Cattle Farming

KAT BAZELEY AND ALASTAIR HAYTON

Foreword by Professor Martin Green

THE CROWOOD PRESS

First published in 2007 by
The Crowood Press Ltd
Ramsbury, Marlborough
Wiltshire SN8 2HR

**www.crowood.com**

**British Library Cataloguing-in-Publication Data**
A catalogue record for this book is available from the British Library.

ISBN 978 186126 975 1

**Acknowledgements**
The most important contributors to this book were the many farmers and
stock-keepers – especially the clients of the Kingfisher Veterinary Practice
and the Farm Animal Practice of the University of Bristol Veterinary School –
who willingly shared their ideas, experience, and enthusiasm. We are
immensely grateful to you all. We much appreciate the constructive advice
and comments on earlier drafts of the text provided by Kathy Anzuino, Colin
Penny, Prof. Patrick Hoffman, David Barrett, James Husband, Richard
Vecqueray, Mike Bray, Ian Powell, Prof. Jon Huxley, Dr Mike Kelly, Stuart
Bacon, and the Milk Development Council. We thank the MDC also for their
kind permission to use material from their Grass+ programme. We are
extremely grateful to Prof. Martin Green for kindly agreeing to write a
foreword. Likewise we should like to thank Peter ('Slasher') Bazeley and
Prof. Gareth Thomas for their enormous efforts in proofreading and large
expenditure on red ink.

Sam Bazeley took many of the photographs for us and deserves special
thanks, as do the other contributors of photographs: Richard Vecqueray,
Ulrike Wood, Lauren Woolley, and the Kingfisher Veterinary Practice. Finally,
many thanks from Alastair to Rachel and their children for putting up with
yet another 'little project', and from Kat to her family for their support.

**Credits**
All photographs by Alastair Hayton, Sam Bazeley and Kat Bazeley, except
where stated otherwise.
Line-drawings by Keith Field and Kat Bazeley.

Edited and designed by
OutHouse!
Shalbourne, Marlborough
Wiltshire SN8 3QJ

Printed and bound in Singapore by Craft Print International Ltd.

# Contents

# Foreword

It is my pleasure to write the foreword for this excellent book. As you would expect from their wealth of experience, the authors Kat Bazeley and Alastair Hayton provide a clear, concise narrative of the fundamental aspects of looking after cattle. Cattle farmers have had to respond to many changes in the last two decades, the result being that many aspects of their work and industry have changed dramatically. This book details these important changes and, in light of them, provides an up-to date view on current best practice for cattle husbandry. The book supplies a practical foundation of knowledge, from basic and essential concepts such as housing systems, to more complex areas, such as the process of an immune response.

The book is comprehensive and well constructed, beginning with an overview of current dairy and beef farming systems, before moving through the major components of cattle husbandry. A particular strength of the book is its focus on current farming circumstances – the opportunities they provide and the constraints they impose. Furthermore, the authors stress the importance of maintaining and improving cattle health and welfare, whilst achieving economically viable production – an essential strategy for cattle farming of today and tomorrow.

We live in a time when the general public are largely divorced from farming and farming methods. Fewer and fewer people have any direct involvement with livestock or any understanding of modern farming practices and circumstances. On the other hand, farmers can lose touch with public opinion and the public mood and therefore become separated from their ultimate market. Such a divide tends to create uncertainty, even suspicion, and books, such as this, that bridge the divide are very welcome. The book will be of great value to those with the intention of entering the cattle industry, student or apprentice, and to the interested observer who wishes to gain an understanding of cattle farming. It also provides useful reminders for those of us currently working with beef and dairy cows. I wholeheartedly commend this book.

*Professor Martin Green, 2007*

# Preface

This book has been written in response to changes in the cattle industry over the last twenty years. Liberalization of international trade has opened up marketing opportunities but has also increased global competition, which, together with loss of government subsidy for farming, has led to reduced profit margins. Cattle farmers have responded by increasing production levels and efficiency of production; this has been achieved by drawing on the information provided by scientific research, particularly in the fields of nutrition and disease control.

At the same time, consumers have become more discerning about the quality of the food they buy, demanding (not unreasonably) that it is safe and free from additives and, increasingly, that farming methods are welfare-friendly. There is also a requirement to ensure sustainable land use and environmental benefits. Farmers must look for ways to meet these disparate targets if they are to be successful.

The answers are not all contained here, but the book provides practical guidance on how to manage cattle well within the context of these various challenges. The text has been condensed to focus on key points, mixing experience from a large number of beef and dairy units with the results of scientific research. Not everyone involved in farming today has come from a farming background, so the fundamentals of the cattle production cycle are explained. Although the authors are both from the United Kingdom, the book concentrates on principles of cattle production that are equally applicable anywhere in the world.

The reader should be left with a sound understanding of the theory and economics of production in the beef and dairy sectors as well as practical husbandry techniques.

# CHAPTER 1

# Dairy Farming Systems

There have been major changes in dairying in the last thirty years, with both average herd size and milk yield increasing significantly: there are now many herds of 500-plus cows producing 12,000–14,000 litres per lactation. These increases in milk production have occurred as a result of changes in the world markets and have been made possible by improved genetic merit as well as better understanding of dairy cow nutrition and computerized ration formulation.

However, increased yield and larger herd size have been accompanied by a variety of problems that cause economic losses, and in

*There are a number of different designs of milking parlour. This rotary parlour is expensive to install but since it enables two operators to milk 150 cows per hour it is well suited to a large herd.*

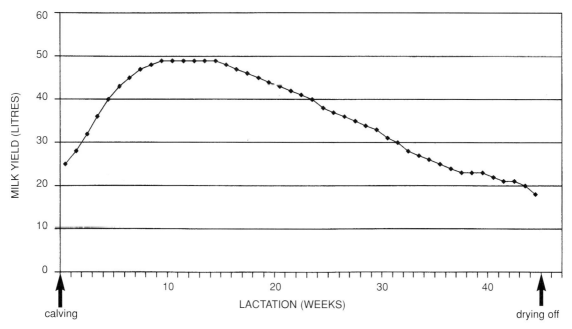

*A typical cow's lactation curve rises to a peak at six to eight weeks after calving; in the following months it gradually falls until weaning or drying off.*

some herds the losses are so high that they outweigh improved performance. For example, despite intensive research there has been no progress in reducing the incidence of mastitis or lameness in dairy herds, with average incidence for each disease running at over 20 per cent (and in some herds at more than 50 per cent). Despite technological advances that should improve fertility management, fertility results are deteriorating, and the number of forced culls is increasing, with many cows culled in their first or second lactation. This represents a great waste of potential and investment.

The challenge to every dairy producer is therefore to achieve the most cost-effective milk production for his or her own unit. This is not necessarily accomplished by producing the highest possible yield, but by making best use of the resources available: farm location, land quality and acreage, housing, skill and availability of labour, genetic potential of the herd, management and capital.

This chapter explores the principles of dairy production and the factors affecting its costs.

## DAIRY SYSTEMS

A number of factors influence herd milk yields, and farmers may choose anything from a low-input–low-output system to a high-input–high-output system. Low-input–low-output herds are almost invariably based on spring-calving grazing systems: they rely on a low purchased-feed rate and try to produce the bulk of their milk from grazed grass. Average yields are around 4,500–6,000 litres per cow per year, so metabolic stress on cows tends to be low, with low incidence of disease and fertility problems, and a low culling rate. At the other extreme, high-input–high-output systems are based on housed cows that are fed high levels of concentrate. They require careful ration formulation and tight fertility management. Yields will be high, ranging from 8,000–12,000 litres per cow per year in the UK, and higher in the United States. Metabolic stress on the cows tends to be high. Either system can work to generate a profitable dairy enterprise, provided that management is good and targets are met. The

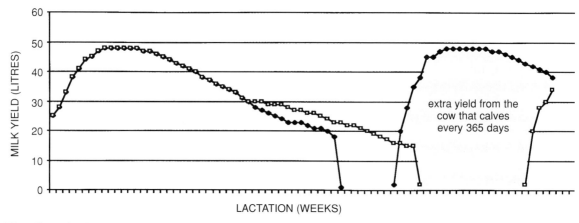

*The effect of calving interval on milk yield. A prolonged calving interval leads to a long period of low production and prolonged dry period. The cow that calves regularly produces significantly more milk per year.*

Milk produced by the cow with prolonged calving interval and extended dry period

Milk produced by the cow calving every 365 days

## MILK PRODUCTION

For the first days after calving the cow produces colostrum to provide her calf with antibodies that protect it from disease (*see also* Chapter 3). The colostrum changes to normal milk by the fourth day after calving. The calf is generally removed from the cow about twenty-four hours after birth (the calf should not be removed earlier because it will not have received as much colostral protection as it needs); in some herds calves remain with cows for longer.

Provided that everything is normal, the cow is ready to join the milking herd within one to four days after calving. Yield increases rapidly, with peak lactation of more than 60 litres daily for high-yielding cows. The transition from dry to peak yield therefore represents a very high metabolic challenge and, not surprisingly, is associated with the highest incidence of disease and weight loss.

Most milk is produced in early lactation, so it is essential that the cow calves regularly (*see* the graph on page 9). It is generally most cost efficient if she calves every 365 days. For very high-yielding cows with very flat and persistent lactation curves there are good arguments that profitability does not suffer if the calving interval is longer than 365 days. This is because the mild reductions in yield will be offset by reduced costs associated with the dry period time and the costs of diseases around calving (periparturient disease). However, at the time of writing there are very few farms that fit this economic model, and the target generally remains to calve annually. But in order to produce a calf every 365 days, the cow must get back into calf within eighty-two days of parturition (because pregnancy lasts for 283 days); at this stage she is still producing a high yield and, unless nutritional management is optimal, may still be losing weight. This represents a great challenge to the dairy farmer.

A cow in a natural environment with little or no metabolic stress can continue to produce a calf regularly and maintain reasonable lactation until she is more than fifteen years old. Her potential milk yield per lactation increases to its maximum at seven to eight years of age. However, the high metabolic demands imposed on the modern dairy cow lead to high culling rates at much younger ages. These culls may be part of a planned strategy for genetic improvement, with cows aged five to six being sold on as dairy animals to make room for replacement heifers with higher potential for milk production. However, many culls are unplanned or forced: cows must leave the herd because they fail to get back in calf or as a result of diseases such as mastitis or lameness.

Cull cows are usually replaced by heifers that may be home-bred or purchased (*see* Chapter 4). Heifers may be reared on the farm or by a contract rearer, and generally calve in at twenty-three to twenty-four months old. Some farmers operate a 'flying herd': they buy in adult cows to be milked for one or more lactations, and then they sell them on and replace them with other adult cows. This saves the cost and labour required for rearing heifer replacements, but there is a risk of buying in disease with purchased cows.

## THE MILKING ROUTINE

The correct protocol for milking cows has an impact on cow health and productivity, and on milk quality. Poor milking routines lead to reduced yields, damage to the cow's teats, transfer of mastitis infection between cows, and contamination of the bulk tank with bacteria.

The basic tenets of a sound routine are:

- A systematic and hygienic preparation of the teats.
- A swift and careful attachment of the teat cups to the udder.
- Removal of the cluster once the udder has been sufficiently milked out (but without over-milking).
- A thorough and prompt disinfection of the teats after cluster removal.

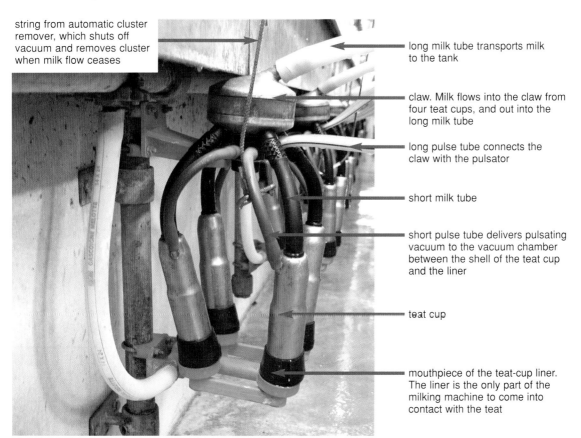

string from automatic cluster remover, which shuts off vacuum and removes cluster when milk flow ceases

long milk tube transports milk to the tank

claw. Milk flows into the claw from four teat cups, and out into the long milk tube

long pulse tube connects the claw with the pulsator

short milk tube

short pulse tube delivers pulsating vacuum to the vacuum chamber between the shell of the teat cup and the liner

teat cup

mouthpiece of the teat-cup liner. The liner is the only part of the milking machine to come into contact with the teat

*The parts of the cluster.*

## The Parlour Routine

### The Milker

Bacteria can survive on skin – and on clothing – for considerable periods. The milker should therefore wear clean, disinfected clothing and latex or rubber gloves. Gloves are especially important as they provide an environment that is less conducive to bacterial colonization than does skin. Gloves should be cleansed and disinfected throughout and after milking to reduce the contamination that occurs during the milking process.

### Timing and Order of Milking

Ideally cows should be milked at the most even interval between milkings, either twice or three times a day. Where possible, high-yielding cows should be milked before the low-yielding cows. Cows with mastitis or a high cell count should be milked with a separate cluster or after all the clean cows.

*Teats are pre-dipped then dry-wiped after thirty seconds. Pre-dipping helps to reduce the spread of environmental organisms at milking.*

DIRECTION OF COWS' MOVEMENT

Second operator attaches teat cups to start milking. This should be done 60–90 seconds after teat preparation to coincide with the cow's milk let-down reflex

First operator prepares cows' teats and uses pre-dip

*The rotary milking parlour in operation.*

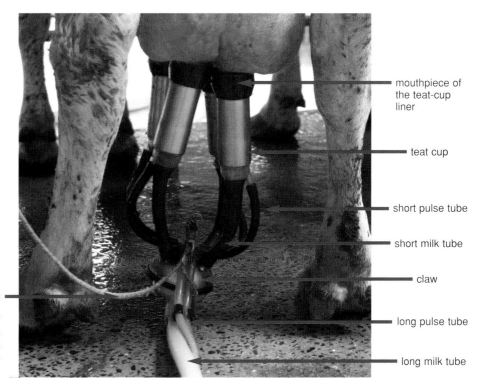

mouthpiece of the teat-cup liner

teat cup

short pulse tube

short milk tube

claw

long pulse tube

long milk tube

string from automatic cluster remover, which shuts off vacuum and removes cluster when milk flow ceases

*The cluster in place on the cow.*

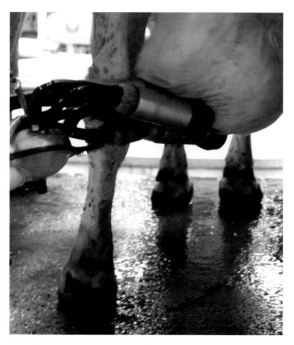

*The cluster being removed by the automatic cluster remover, which has registered cessation of milk flow.*

### Pre-Milking Routine

A variety of different practices are used in the preparation of the cow's teats prior to the cluster attachment. This varies between farms and within farms, depending on general teat cleanliness and the levels of mastitis currently experienced by the farm. The following should be considered best practice:

- Wash dirty teats with water containing a disinfectant and dry thoroughly with individual towels.
- Examine/palpate udder and foremilk onto the parlour floor.
- Pre-dip the teats with a proprietary pre-milking dip.
- Leave for thirty seconds and then dry teats with an individual towel for each quarter.

Pre-milking teat disinfection and fore-milking are both useful procedures and best practice, but they may not be essential on farms with low levels of mastitis and bacterial contami-

nation of the bulk tank, provided an alternative method of mastitis detection is employed.

### Milking

The cups should be attached to the teats within sixty to ninety seconds of the teats initially being handled because it is important to ensure that the attachment of the cluster coincides with the cow's milk let-down reflex (mediated by the hormone oxytocin). This allows a fast and efficient milking out.

Care should be taken to ensure that the cups are level and balanced on the udder, thereby preventing air ingress at the mouth of the liner, and to ensure a full milking out of all four teats. Once the cow is milked out the cluster should be removed. This is usually performed by the automatic cluster removers (ACRs). The vacuum should be shut off prior to cluster removal and the vacuum equilibrated so that the cluster falls off the teat in an easy motion and does not require to be pulled off.

### Post-Milking

Following the removal of the cluster the teats should be treated as soon as possible with a proprietary post-milking teat disinfectant. This is a vital stage in the milking routine, as a failure to disinfect the teat correctly – by not ensuring the whole teat skin is covered in disinfectant – will result in the potential transfer of mastitis-causing bacteria between cows, thereby spreading disease. The disinfectant can be applied with a spray or by dipping the teats. Either method is acceptable provided it is done thoroughly and cleanly. As a guide to quantity required, a teat dip will use 10ml per cow per milking, and a teat spray will use 15ml per cow per milking.

### Cleaning of the Plant

Post-milking, the plant requires disinfection to prevent the build-up of bacteria that can adversely affect milk quality. A variety of different systems are employed in the process of plant cleaning. It is essential to ensure that the plant's internal cleaning systems are functioning correctly. For example, the water heater

*Cows in the collecting yard, waiting to be milked, are pushed forward by an electronic gate. This reduces the labour required to move cows into the parlour for milking.*

must attain the correct temperature and have sufficient capacity for the plant; any directions on chemical use must be followed precisely.

## The Milking Machine

An incorrectly functioning machine has the capacity to damage teats and drive infection into the udders. Obviously both are undesirable, and ensuring the machine is well maintained and checked regularly for correct function is a vital component of mastitis management. Machines should be checked and serviced at least twice a year.

Teat-cup liners are prone to wear and should be changed regularly to prevent them from becoming harmful. Rubber liners should be changed every 2,500 milkings or every six months, whichever measure is the shortest.

## Collecting and Dispersal Yards

Cows may have to wait for considerable periods to be milked, so they should have adequate space (1.5sq.m per animal) in the collecting yard, and water should be provided. A gradient of 3–5 per cent up to the parlour will encourage cows to orientate themselves towards the parlour and assist drainage when the yard is cleaned. Heat stress should be avoided by the provision of good ventilation via a ridge or slotted roof and the use of sprays and fans to cool the cows.

The exit route after milking should not include sharp turns that will slow the flow of cows and may damage feet. Handling facilities for treatment should be readily accessible from the parlour, as well as a footbath. Water should be available. Following the exit from the parlour cows should stand on a clean concrete surface for at least thirty minutes before they are given access to the bedded area. This allows time for the muscular teat sphincter to contract, sealing the teat from the outside environment, before the cow lies down, which reduces the risk of bacteria in the bedding penetrating the teat and the udder.

# ECONOMICS OF DAIRY PRODUCTION

The profit from any given enterprise is calculated as the total output less the costs of production. The costs of production are usually split between variable and fixed costs.

The fixed costs of production relate to:

- Labour (paid and family input).
- Machinery and power.
- Rent and property costs.
- Finance.
- Sundries.

The variable costs include:

- Feed (purchased and home grown).
- Veterinary attention and medicine.
- AI and milk recording.
- Other (e.g. bedding).

## Fixed Costs and Economies of Scale

As their name suggests, fixed costs for a farm tend not to vary with output whereas variable costs tend to relate directly to the level of output. Therefore the more litres that are produced by the farm the less the fixed costs contribute as a percentage of the total cost of production of each litre. Thus, the profit for each litre should be higher. This observation, described as the 'economies of scale', has driven many farmers to increase herd size and yield.

For example, if the fixed costs of production for 100 cows are £50,000 per annum, and the price for each litre is 18 pence, if the cows yield 6,000 litres per annum the percentage of fixed cost for each litre of milk will be:

$$\frac{50,000}{(6,000 \times 0.18)} = 46\% \text{ of cost of production.}$$

If, however, the farm produces 10,000 litres per cow per annum the percentage cost of production due to fixed costs will be:

$$\frac{50,000}{(10,000 \times 0.18)} = 28\% \text{ of cost of production.}$$

This example is too simplistic as there are a number of fixed costs that will vary with output, e.g. electricity for the parlour and milk

---

**Table 1: Organic versus conventional dairy margins**

|  | Organic | Conventional |
|---|---|---|
| Yield per cow (litres) | 6,000 | 7,000 |
| Milk price (pence per litre) | 29 | 18 |
| Concentrate (kg per litre) | 0.22 | 0.25 |
| Concentrate (pounds per tonne) | 250 | 125 |
| Feed cost (pence per litre) | 5.5 | 3.1 |
| Stocking rate (cows per hectare) | 1.71 | 2.26 |
| Margin over purchased feed – per litre (pence) | 23.5 | 14.9 |
| Margin over purchased feed – per cow (pounds) | 1,410 | 1,043 |
| Margin over purchased feed – per hectare (pounds) | 2,411 | 2,357 |

In this example it is easy to see that the organic system commands a much better profit margin over purchased feed per litre and per cow, but because the stocking rate is much lower the overall margin over purchased feed per hectare is very similar.

**Table 2: Comparison of different production systems on margins**

|  | Low-input–low-output | High-input–high-output |
|---|---|---|
| Yield per cow (litres) | 5,000 | 10,000 |
| Milk priced (pence per litre) | 18 | 18 |
| Stocking rate (cows per hectare) | 2.0 | 3.0 |
| Net margin – per litre (pence) | 3.0 | 1.0 |
| Net margin – per cow (pounds) | 150 | 100 |
| Net margin – per hectare (pounds) | 300 | 300 |

cooling; labour and machinery input with an increased herd size. Furthermore, when assessing the impact of increasing output on cost there needs to be a distinction between increasing milk output through higher yield per cow and increasing it through increased herd size. Increasing milk yield per cow will have a small impact on fixed costs (e.g. additional milk cooling), whereas increasing herd size could have a very large impact on fixed costs, including the need for additional buildings and equipment. These factors can offset any potential increase in profit unless the proposed expansion is very well planned and well managed.

When considering increasing outputs a major limitation for every farm business will be the existing infrastructure available (parlour, bulk tank, housing, feed storage, slurry storage), due to the high cost of providing an additional cow place. Land is usually not a limiting factor (except in organic systems) as it can either be rented or purchased as feed.

**Overall Farm Profitability**

As already stated, aside from any personal beliefs on the appropriateness of a particular farming method, the best system for the farm is not the one that will maximize the profit per litre but the one that maximizes the potential profitability of the farm as a whole.

A good way to illustrate this is to examine the effects on a farm when it is farmed conventionally and the effects when it is farmed under organic standards. Organic farms are restricted in the type and level of land inputs (fertilizer/herbicides) they can use, with a con-

sequent reduction in the level of home-produced feed they can produce. The inevitable result of this is a reduction in the number of cows the farm can feed. They are also restricted in the level of concentrates they can feed, which will obviously impact on the potential yield a cow can sustainably achieve. The positive side of the equation for the farmer is that organic milk commands a much better price per litre than conventional milk.

If we examine a financial model of a farm when farmed under the two systems it is possible to see how these variations in profitability per litre produced and overall farm output relate (*see* Table 1, opposite).

Low-input systems rely on maintaining very low purchased-feed input to maximize profitability per litre produced. As such they rely primarily on home-grown forage (especially grazing) to feed the cows. Similarly, given the restrictions on concentrate use and the cost of purchased organic feed, organic farms rely very heavily on producing their own feed. The result of these low inputs is low outputs and a reduced stocking density, as compared to a farm prepared to buy a large amount of feed from outside the farm. Low-input and organic farms' profitability is therefore very dependent on stocking rate, which in turn is a reflection of the quality, productivity and efficiency of use of their forage. High-input farms are less sensitive to changes in their forage utilization (as stated earlier, high-yield systems can provide total control over feed supply, which reduces risk and variability) but instead are more sensitive to changes in purchased feed price.

---

## Summary of Sales Income and Variable Costs

### Sales Income

1. Price paid per litre milk:
   - Milk buyer contract.
   - Organic vs conventional.
   - Volume and transport.
   - Fat and protein %.
   - Seasonality.
   - Farm assurance.
   - Milk hygiene.
   - Market-related bonuses.

2. Number of litres of milk:
   - Number of cows.
   - Lactation curve and stage of lactation.
   - Age of cows.
   - Fertility.
   - Nutrition.
   - Disease.
   - Breed.
   - Genetic potential.

3. Cull cows (number and price per animal):
   - Number – culling rate.
   - Price.
   - Forced vs planned culls (e.g. lame cull cows have no value).
   - Market variation.
   - Body condition score.
   - Weight.

4. Calves (number and price per animal):
   - Fertility of adult herd.
   - Mortality of calves.
   - Breed and genetic potential.
   - Age.
   - Sex.
   - Market variation.

### Variable Costs

1. Purchased feed (quantity and quality) – closely linked with yield:
   - Quantity (including waste).
   - Price (£ per tonne).
   - Organic versus conventional.
   - Compound versus straights.

2. Forage costs:
   - Cost of seed, fertilizer and sprays.
   - Forage yield.

3. Replacements:
   - Number.
   - Cost (purchased or home reared).

4. AI or bull:
   - Fertility – services per conception (number of straws or number of bulls to get cows in calf).
   - Cost per AI straw.
   - Purchase price, less sale price, of bull(s).
   - Culling rate of bulls.
   - Maintenance cost of bulls (number × cost per day).

5. Veterinary and medicine costs:
   - Disease incidence.
   - Costs of medicines.
   - Preventive veterinary care.
   - Discarded milk.
   - Variable costs do not include labour, but labour costs are significant for disease.

6. Bedding and sundries.

---

## Sales Income and Variable Costs

Since it is difficult to alter fixed costs, it is necessary to understand the factors that affect sales income and variable costs in order to improve herd profitability.

## Sales Income:

*Price Paid per Litre of Milk:*
The milk price is influenced by a number of external factors outside the control of the farmer, such as world supply and demand, EU support system, and local supply and demand. The price paid for the milk is determined by the choice of milk buyer and the milk contract available in a particular location. Other factors that influence price are:

- Organic versus conventional production. In the UK about 2 per cent of milk produced is organic. Yields per cow are lower but organic milk usually attracts a premium price.
- Seasonality and profile. Processors generally want an even supply of milk into the

*Low-input–low-output system makes best use of spring pasture, with cows calving in spring so that peak yields coincide with lush growth of high-quality grass.*

factory. To encourage this, prices for milk will be set on a seasonal basis to encourage production in the October/November (lowest UK production) and to discourage production in April/May (highest UK production). In addition, some liquid buyers pay a premium for a level profile of production.

• Milk constituent quality. The price paid for a particular level of milk fat or protein depends on the intended use of the milk. Milk protein is not particularly important for liquid consumption and usually no premium is paid; where fat can be separated to produce semi-skimmed and skimmed milk, a fat payment is usually paid which relates to the market value of cream. For manufacturing (butter, cheese and powder), the processor normally pays on the percentage of fat and protein in the milk to encourage a greater supply of the constituents that are used in the product.

• Farm assurance. Within both conventional and organic milk pools there are a variety of assurance schemes looking to safeguard the safety and integrity of the production process. Not all schemes (if any) attract a premium for the milk, but failure to register can severely limit access to buyers.

*The Jersey cow produces a lower yield of milk of high quality than do other breeds. They are small cows (approximately 400kg).*

- Milk hygiene. Milk hygiene relates to the levels of bacteria and cells within the milk. Consumers are increasingly aware of – and concerned about – the provenance and quality of food. Many processors demand high-quality milk for its better keeping qualities and higher yields of cheese. Most processors set price bands to discourage the production of milk with high cell counts and bacterial load.
- Volume and transport. Most milk buyers will pay a bonus relating to the volume of milk collected and/or charge a collection

**Table 3: Comparison of different dairy breeds**

| Breed | Yield | % Fat | % Protein |
|-------|-------|-------|-----------|
| Holstein | 6 | 1 | 1 |
| Holstein-Friesian | 5 | 2 | 2 |
| Ayrshire | 4 | 4 | 5 |
| Shorthorn | 3 | 3 | 3 |
| Guernsey | 2 | 5 | 6 |
| Jersey | 1 | 6 | 4 |

(6 = Highest  1 = Lowest)

*The Holstein cow produces highest yields with lower fat and protein content. They are large cattle, so their feed requirements are high. They are best suited to high-input–high-output production systems, which optimize their yield potential.*

fee. This reflects the high cost of collecting milk from individual farms, especially where the volumes of milk to be collected are small. There may also be a bonus for every-other-day collection to encourage larger volume collections.
- Market-related bonuses. There are a number of bonuses that relate to particular markets, especially in the 'supermarket' liquid sector. These pay a premium for milk dedicated to a particular milk pool.

*Number of Litres of Milk*

This depends on a number of factors including:

- Breed. Breed of cow has a major influence on yield (*see* Table 3). Cows of different breeds may be more suitable to one dairy system than to another; in the wrong system, they may be more prone to disease, so yields may be lower than predicted.
- Genetics. High-genetic-potential cows tend to peak higher, reach peak yield later, and

21

show more persistency (a flatter lactation curve).

- Age. Heifers produce less milk at peak lactation than do older cows (first-calf heifers should peak within 25 per cent of older cows), but their yield declines more slowly. After peak lactation, heifers will drop about 0.2 per cent milk per day, and older cows will drop about 0.3 per cent milk per day (or 3 per cent every ten days). Cow milk production per lactation increases with succeeding lactations until the cow is seven to eight years old.
- Month of calving. This principally relates to the effect of housing and turnout, which is really the effect of feeding. For example, cows turned out to plentiful pasture in mid- to late lactation may maintain yields for longer.
- Frequency of milking. Three times daily milking tends to give higher peaks and greater persistency.
- Feed level. Provided the cow is fed at the correct feed level (energy and protein) for the stage of lactation then the normal level of fat and protein for the stage of lactation will be produced. Invariably, if the correct feed level is not fed then milk quality will suffer owing to a lack of nutrient supply. This is often seen during poor grazing conditions (too wet or too dry).
- Nutritional system. A flat-rate, total mixed ration (TMR) system produces a much flatter lactation curve than a low-input, spring-grazing system with little extra concentrate. (*See* Chapter 5.)
- Nutritional deficiency or disease in early lactation. If this results in reduced production at peak, production throughout the lactation will be reduced. The basic rule of thumb is that for each extra kilogram of milk at peak, a cow will produce 200–225kg more milk over the entire lactation.
- Chronic mastitis. Chronic mastitis causes permanent damage to the milk-producing tissue in the udder, so potential for milk production is reduced even if the infection is eliminated. Damage is often observable

as 'light' quarters or hard, scarred tissue in the udder.
- Fertility. Yield per cow is maximized if average calving interval is 365 days. Longer calving intervals result in a longer dry period and/or more cows in late lactation producing low yields, so overall milk production is reduced.
- Other disease. Disease at any stage of lactation can affect milk yield and fertility.

## Cull Cows and Bulls (Number and Price per Animal)

The number of cows sold as culls depends upon the culling rate, which may vary between 15 and 40 per cent. Very low culling rates (less than 15 per cent) may mean there are many old, unproductive animals in the herd, while a very high culling rate generally indicates wasted potential. The price paid per animal depends upon age, weight and body condition score, and there may be considerable market variation according to demand. Thus, planned culls generally attract a higher price than forced culls because the farmer can choose when to sell the animals. Cows or bulls that must be culled for lameness or because they are down (unable to stand) have no economic value because they cannot be transported from the farm. There will be a cost to the farm for the removal of animals that die (fallen stock).

## Calves (Number and Price per Animal)

The number of calves available for sale depends upon the fertility of the adult herd, the replacement rate (which determines the number that are surplus to requirements), and the mortality rate of calves (*see* Chapter 3). The price paid per calf depends in turn upon:

- Breed and genetic potential. Beef-cross calves tend to attract higher prices.
- Age. Older calves are usually more valuable, but this must be offset against extra costs and labour to keep and feed calves for extra time.
- Sex. The highest value will generally be for male beef calves, followed by female beef

calves, with the lowest value for pure dairy-bred male calves. This reflects their value for beef finishing. Sexed semen is now being used as this lowers the proportion of low-value dairy-bred male calves.

- Disease. Any disease that reduces growth rate is likely to reduce the value of calves.
- Market variation (for example, availability of export trade). These factors are discussed in detail in Chapter 2.

## Variable Costs

### Feed (Quantity and Quality)
The diet cost is closely linked with yield, ranging from low-input–low-output to high-input–high-output.

- Ability to grow feed. The ability to grow different forages and straights (*see* Chapter 5), and the quantity of forage that the farm can produce, has a major influence on feed cost. Grazing is the cheapest forage and, on farms where there is plenty of high-quality grazing, feed costs can be minimized. Maize and cereal whole-crop silage are usually produced at a lower cost than grass silage as they are a one-cut harvest; grass silage is made from several cuts, with the later cuts being expensive per tonne of dry matter, especially where contractors are used.
- Organic versus conventional. Costs of producing organic forages are higher than conventional, and organic concentrate feed is also significantly more costly than conventional concentrate.
- Concentrate (compounded feed) versus straights (single feed components). Concentrates are more expensive than the combined cost of individual constituent feeds. However, straights are normally fed out of parlour; storage facilities and a mixer wagon should be available; and labour must be available to mix the feed.
- Wasted feed. This may be significant on some farms where there is spoilage at the silage face or there is poor wrapping of large bales.

### Replacements
The cost of rearing replacement heifers depends upon the number required, which in turn is determined by culling rate, calving interval, mortality, and age at calving (*see* Chapter 4). The costs per day per animal include nutrition and incidence of disease. The number of days required for each animal depends upon age at calving, which in turn varies with growth rate and incidence of disease.

### AI or Bull
The most important determinant of these costs is the fertility rate of the herd, measured as number of services per conception (*see* Chapter 7). This determines the number of AI straws required or the number of bulls to get cows in calf. Other factors are:

- Cost per straw. This may vary from very cheap to very costly. High-cost straws are those from popular, well-marketed bulls, and there are plenty of high-genetic-merit bulls available to improve the genetic merit of the herd. Where fertility results are poor (more than three services per conception) it may be difficult to justify the use of high-cost straws.
- Purchase price of bull(s). Higher-value bulls generally have higher genetic merit for producing valuable offspring, so it can be cost effective to buy an expensive bull. High-genetic-merit herds will normally use AI, whereas lower genetic merit, grazing-based systems will usually use a range of bulls for crossbreeding.
- Culling rate of bulls. This is in part a function of the breeding plan, since if home-bred heifer replacements are to be kept, the bull must be replaced every two years to prevent inbreeding. Culling rate also depends upon the housing, feeding and management conditions of the bulls, and is likely to be lower if bulls are rested regularly, kept in a straw yard or at pasture, and not allowed to become excessively fat.
- Maintenance cost of bulls.

*Veterinary and Medicine Costs*

The effects of disease can be crippling on a herd's financial performance. The majority of these losses are not due to the more obvious costs of veterinary time, medicine costs and milk discarded in observing medicine withdrawal periods; rather they arise from a loss of yield and increased culling. For example, veterinary time and medicines account for only 18 per cent of direct costs and 7 per cent of total costs of a lameness case; the remainder is due to reduced milk yield and fertility, and an increased risk of culling. Therefore a high spend on veterinary time and medicine may be justified if it is proactively spent to prevent disease rather than reactively to treat it. In general, veterinary and medicine costs tend to increase with yield, but the cost per litre is often similar.

*Bedding and Sundries*

The bedding cost varies enormously according to the housing and bedding system used. Loose yards can result in a very high bedding cost, with around 3 tonnes per cow of straw being used in a 200-day winter. The straw requirement in a cubicle is around 1.4 tonnes per cow, which reduces to 1 tonne per cow if chopped and to 0.6 tonnes where a mat or mattress is used.

## CHOICE OF SYSTEM FOR A PARTICULAR UNIT

It is clear from the above that a multitude of factors contribute to the overall profitability of a dairy unit. Many are interlinked, and the dairy farmer must consider them all, separately and together, when choosing the most appropriate management system. Most important is to make sure that, whatever system is chosen, it can be done well, so that losses due to poor fertility, disease and/or high numbers of forced culls can be kept to a minimum.

A number of practical considerations must be taken into account:

- The quantity and type of forages available. This will be influenced by farm location, soil type and forage-yield potential. A farm that has plenty of grazing can optimize its use with a spring-calving, low-input–low-output system or by producing high yields of conserved forage. If little grazing is available and stocking rates must be high, a more intensive feeding programme for high yields is appropriate.
- Type of feeding system available to the farm. Out-of-parlour feeders or mixer wagons are required to provide a balanced ration for high yields. This involves a higher capital and running cost, which will need to be spread over a high output of milk.
- Infrastructure (housing, feed storage, milking parlour, slurry storage, field access). The infrastructure available will have a major impact on the choice of system (owing to the high capital cost of investing in additional facilities). High-yielding cows require high-quality housing to prevent disease problems. Group management systems work only if separate groups can be fed, milked and housed easily.
- Farm personnel strengths and weaknesses. The skills, level of knowledge and time available should be analysed for available staff, and a system chosen that makes best use of them.
- Genetics of the herd. The cows should be suited to the feeding system. For example, high-genetic-merit cows have been bred to produce large quantities of milk, and if a diet fails to meet their requirements, they lose excess weight, with consequent increase in disease and reduction in fertility. Low-input, grass-based systems or organic production are unsuitable for such cows.
- Calving pattern of the herd. The market and milk buyer requirements will have a major bearing on the calving pattern. Low-input systems that rely on grazing require cows to calve in a tight block in spring so that their maximum feed requirements coincide with plentiful grass at its best quality.
- Storage and handling of slurry. Farmers are increasingly required to ensure that they do not cause environmental damage

by inappropriate handling of slurry. Slurry handling systems can be expensive, so the ability to handle and store slurry safely may impact on choice of system.

The choice, once made, should not be set in stone. The system should be reviewed regularly according to experience and changing circumstances, and altered as needed.

## FARM RECORDING SYSTEMS

Farm records should not only cover financial and production data for the herd but also the causes of good or bad performance. Using the example of motor racing, it is no use to a racing team to know that a car is going slower than expected if they have no means of determining why, and consequently such teams invest a vast amount of time and money in developing such systems.

Good recording systems should be easy to use, avoid excessive duplication in the capturing of data, and detailed enough to permit suitable analyses to be performed. They can be used to detect problems and identify their causes, and, once management solutions have been applied, to monitor the outcome of

*Assessing the calving pattern of a herd using a bray board. Each cow is represented by a numbered, magnetic marker, which is placed on the board according to her calving date. This provides a visual record of the herd's fertility status. (Photo: Kingfisher Veterinary Practice.)*

changes to ensure that success has been achieved.

This chapter has introduced a daunting range of issues that impact on dairy production. These are explored in more detail in the remaining chapters.

# CHAPTER 2

# Beef Farming Systems

The aim for all beef farmers is to sell beef animals at a higher price than they have cost to produce. To understand how this can best be achieved, we need to examine the market and the pressures that determine those prices and costs. Liberalization of trade both globally and within the expanding European Union means that farmers must compete in a world market, where it is predicted that global demand for beef will rise, but where production costs vary

*Well-grown beef heifers housed in a loose yard on straw bedding. These are dairy-cross animals bought in for fattening on a beef-rearing unit.*

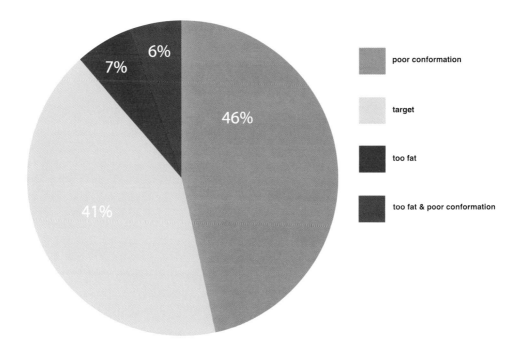

*Pie chart illustrating the percentage of prime beef in main classification sectors, with many cattle failing to meet the conformation or fat cover required by retail outlets. (From EBLEX 2005.)*

enormously according to climate, labour costs and regulation. Whatever the production system, profitability will stem from efficient production of carcasses of the correct weight and quality to meet consumer requirements.

Although there is a growing market for direct sales of beef through farm shops and farmers' markets, beef is mostly retailed through supermarkets and through independent butchers. Abattoirs that supply these retail outlets require carcasses of consistent size, conformation and level of fat cover, and their price grid reflects this, with premium prices (up to 8p per kilogram in the UK above base price at the time of writing) paid for the best, and much lower prices (up to 30–40p per kilogram less than base price) for carcasses that are, for example, too fat and/or of poor conformation. Currently only 54 per cent of UK carcasses fall into the categories that attract the best prices, so there is room for improvement.

In this chapter, the key features of the industry and the factors that affect the effi-

ciency and potential profitability of the beef herd are examined.

## BASIC PRINCIPLES OF PRODUCTION

Beef farming systems are remarkably diverse, ranging from suckler herds where home-bred calves are reared to sell as finished beef, to herds where cattle are purchased at any age from a week old, to be sold at any age from a few weeks to slaughter weight. There is also the pedigree beef industry, which produces breeding bulls and cows or heifers. Rearing methods are equally diverse: from extensive grazing on permanent pasture throughout the year to heavy cereal feeding of housed beef cattle to achieve liveweight gains of up to 1.4kg per day.

The aim is to seek the production level that maximizes economic return by adopting the system that best suits the farm. Factors that should be considered in choosing a suitable system include:

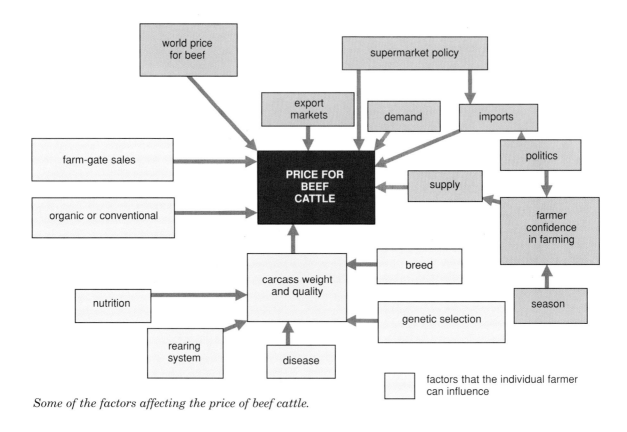

*Some of the factors affecting the price of beef cattle.*

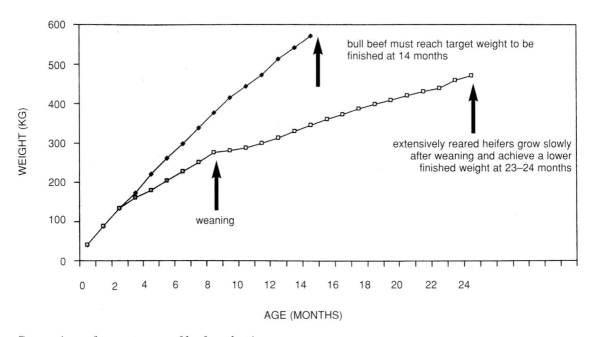

*Comparison of two extremes of beef production.*

- The capacity of the land to support growth of grass, forage and cereals for cattle feed, and the likely quality of pasture (*see also* Chapter 6).
- Land area available.
- Winter housing available.
- Labour (skill and number of stockpersons).
- Availability of cheap straw bedding.
- Other enterprises such as sheep-rearing and arable cropping.

Overall efficiency of production is determined by the fertility of the breeding herd (discussed later), the quality of carcasses produced, and the animals' growth.

## Carcass Quality

The payment a farmer receives is based on the weight, conformation and fat class of the carcass. Conformation will largely be determined by the genetics of the animals. The level of fat has a sex and genetic link but it is heavily influenced by the diet fed. For any given class of fatness there will be a range of possible slaughter weights. Beef breeds are split into two basic breed types: either early-maturing breeds (such as the Aberdeen Angus) or late-maturing breeds (usually Continental breeds, such as the Limousin or Charolais). The term 'maturing' refers to the likelihood of fattening in response to increasing energy density of the diet, with early maturers disposed to becoming fat more quickly than late maturers for any given energy density of diet.

The problem with fattening too quickly is that the animal reaches target fatness before it is large enough to meet the market requirements. This phenomenon is also influenced by the sex of the animal. Bulls are late maturing, steers can be considered intermediate, and heifers are early maturing. The feed rate should therefore match both the breed and the sex of the animal. Heifers and native British breeds are unsuitable for intensive feeding programmes because they fatten too early. This does not mean that only one breed, or cross, or sex, should be used within a system,

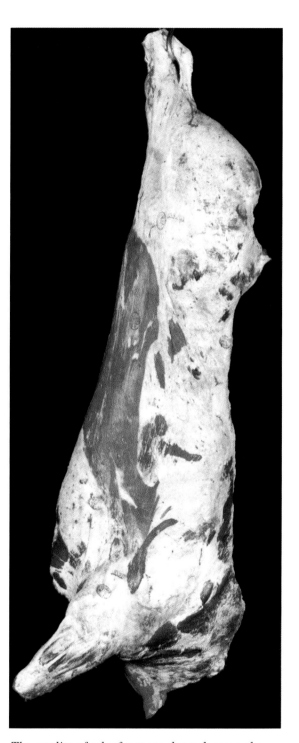

*The quality of a beef carcass depends upon the ratio of meat to bone, and the fat cover both around and within the meat (so-called marbling). (Photo: EBLEX.)*

29

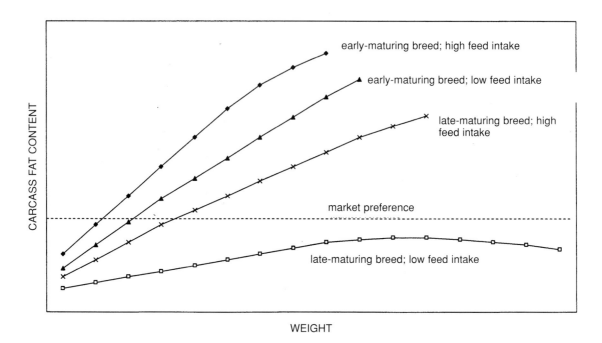

*Graph showing variation in carcass fat content and weight with feed energy density and breed. (Based on Allen, D.,* Planned Beef Production, *1990).*

but that they should be managed accordingly, using separate penning and feeding.

Earliness of maturation is of particular importance in feeding systems that rely on finishing at grass. This creates a time limit in that if the breed matures too slowly, there is a danger that the cattle will fail to mature before the end of the grazing period and require expensive supplementation to finish.

## Growth Rates

Although growth rate is an important determinant of beef production efficiency, the most cost-effective rearing method for a particular unit may not be the one that achieves highest growth rates. Growth rate is influenced by a number of factors:

1. Nutrition.
   Nutrition is the most important determinant of growth rate, and accurate ration formulation for energy, protein, minerals and trace elements is important. The growing animal must satisfy its needs for both maintenance and growth. Rapid growth rates therefore tend to give more efficient use of feed (because maintenance costs to reach the same end weight are lower). Younger cattle require and make best use of highest-quality feed because the efficiency with which they convert feed is higher than that of older animals. The quality of pasture changes throughout the grazing season, so it may be necessary to balance grass with other feed to optimize growth rate. For example, the high protein content of spring pasture must be balanced by adding cereal, and it may be impossible to fatten/finish cattle at pasture in late summer; instead they should be brought in and finished on a ration of forage and concentrate. In order to make best use of available pasture and forage, cattle may be fed to achieve slow to moderate growth rates (0.4–0.6kg per day) over the winter period, and rely on a compensatory surge of growth when they are turned out to pasture in spring. This period of restricted

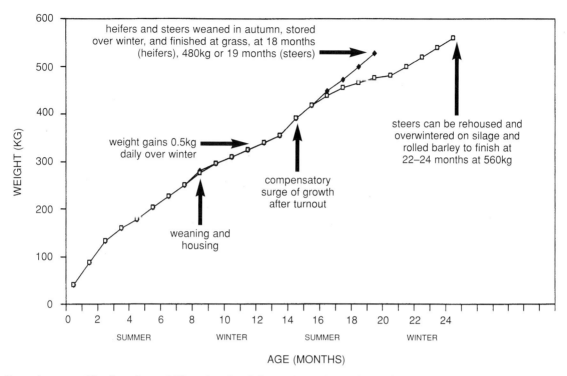

*Growth rates of beef cattle on hill and upland farms in spring-calving herds. Calves grow rapidly, suckling at grass in the first summer. Thery are weaned in autumn, stored over winter, and either finish after a second summer at grass or over the following winter at two years old.*

growth is called the store period. Over-feeding cattle during the winter increases feed costs and reduces liveweight gain at grass but, if cattle are to be finished at grass, it will be necessary to ensure sufficient gain during winter so that the animal has a chance of finishing at pasture.

2. Breed and genetic potential.

Growth rate is a heritable trait with some breeds having a much higher potential for rapid growth than others. Selection for growth rate and conformation, using the Estimated Breeding Value system (*see* Chapter 7) has led to large increases in genetic potential for growth rates in the last thirty years. There may also be the opportunity to exploit hybrid vigour by crossbreeding. Hybrid vigour is the phenomenon whereby crossbred animals show higher genetic potential (such as for growth) than either of the purebred parents. The extra cost of buying in cattle of beef breeds with highest potential for growth may offset higher sale prices unless their potential can be met within the rearing system (housing, nutrition and management inputs) available. The conformation of some breeds may not be appropriate for high feed and growth rates.

3. Sex.

Bulls have the highest potential for rapid growth (at least 1.2kg daily for late-maturing breeds, 1.0kg daily for early-maturing types), with a feed conversion efficiency 10–12 per cent higher than for steers. They produce the largest carcasses with least fat cover. Steers have a higher potential for growth than heifers, who tend to finish at lighter weights, with lower average weight gains. In some

*This gate has been bent by bull beef animals, illustrating why pen construction must be robust.*

countries (though not the EU) hormones are used to increase growth rates of heifers and steers. Bull beef rearing is generally suitable only if cattle can be permanently and securely housed in groups of not more than twenty, with excellent handling facilities and adequate safety arrangements for stockpersons. Bulls can be reared at pasture, provided there is no risk to the public: fencing must be secure, they should be away from populous areas, and prominent warning signs should be displayed. The target is that bulls should achieve a high finished weight at 380 days. Older bulls, particularly if they fail to meet target carcass weight of 275kg, incur a severe price penalty. A recent UK survey found that results for Holstein-Friesian bulls were very variable. Bull beef cattle are usually fed ad lib on a concentrated cereal ration with straw, or on a complete ration including a high proportion of maize silage.

The diet poses a risk of acidosis unless very well managed (*see* Chapter 5).

Some farmers delay the castration of calves for several months in order to optimize early growth rates. However, late castration may lead to a growth check and carries a higher risk of post-operative complications such as infection and haemorrhage.

4. Disease.
   Almost all disease has some effect on growth rate, but the most marked effect is seen from scouring, pneumonia and gut parasites. It is therefore important to institute preventive measures that reduce or eliminate these diseases, and to treat early any disease that emerges.

5. Mineral/vitamin/trace element status.
   Growing cattle have high requirements for trace elements such as copper, vitamin E,

selenium, iodine and cobalt. If these are not met, growth rate may be compromised, even where there are no clinical signs of deficiency. It is important that the balance between minerals such as calcium and phosphorus is correct, or abnormal bone growth may occur.

# PREPARING CATTLE FOR SLAUGHTER

Cattle being prepared for sale or slaughter must comply with a series of regulations. A high cost is incurred if cattle must be kept back at the farm for extra time, or if they are rejected from the market or abattoir because they fail to meet standards. The regulations include:

1. Cattle cleanliness.
   Dirty cattle are no longer accepted at the abattoir because of the risk of contaminating carcasses with disease organisms that can spread to humans (for example, salmonella and *E. coli* 0157). In order to keep housed cattle clean you should:

   • Ensure that stocking density is not too high.

• Provide plenty of straw bedding.
• Scrape and bed up regularly.
• Provide a diet that contains adequate fibre so that dung is firm.
• Control parasites (worms and fluke).

If cattle require cleaning up before slaughter they should be moved to a marketing pen where stocking density is low and clean straw bedding is plentiful – it may take three to four weeks if coats are very dirty. Cattle can also be clipped immediately prior to transport to remove dirt. (*See also*, FSA 2004, *Clean Beef Cattle for Slaughter: A Guide for Producers.*)

2. Identification and passports.
   Cattle cannot be sold in the UK without two eartags and a passport. Calves must be tagged and registered before they are twenty-seven days old, and no late registration is allowed. This means that a calf

*Dirty cattle like this cannot be sent for slaughter. They must first be cleaned up in a marketing pen, which can take several weeks.*

that is not registered on time can never leave the farm, even direct for slaughter. Beef cattle commonly lose eartags, particularly when housed, so these must be regularly checked.

3. Withdrawal times for drugs.
   All drugs carry a meat withdrawal period (which may be zero), and their use must be recorded in a medicines book, and withdrawal times observed. The farmer must consider potential sale or slaughter dates when using any treatment or preventative medicine; withdrawal times of certain long-acting treatments for parasites (bolus anthelmintics) may be very prolonged. If cattle become sick shortly before they are due to be sold, treatments should be chosen that are likely to be effective but have a short withdrawal period.

4. Six-day standstill.
   No cattle or sheep can be moved off a holding within six days of their arrival on the holding, except to slaughter and for certain specified exemptions (*see* DEFRA website for current regulations). This includes cattle that have been grazing on a different holding, away from the main unit.

5. Pre-movement testing for TB.
   At the time of writing, all cattle over six weeks old from holdings on annual or two-yearly TB-testing must be tuberculin-tested within sixty days before sale.

6. Vehicle Cleanliness.
   Vehicles used for transporting cattle must be thoroughly cleaned both before and after transport, using a DEFRA approved disinfectant (Transport of Animals (Cleansing and Disinfection) (England) (No.3) Order 2003).

7. Journey plan.
   Ideally cattle should be transported for as short a distance as possible, but the Welfare of Animals (Transport) Order 1997 specifies that journeys must not be longer than eight hours, or, if more than eight hours, must be carried out by transporters holding a Specific Authorization (to carry livestock on journeys of over eight hours by road) in a vehicle that meets a strict set of criteria.

Each factor should be checked well before the planned sale date to make sure that cattle comply with current regulations. The details are likely to change with time, so should be confirmed with the relevant authority.

# THE STRUCTURE OF THE UK BEEF INDUSTRY

There are, broadly speaking, three main categories of beef unit:

1. The beef rearer.
2. The suckler herd.
3. The pedigree herd.

The links between these three types of unit and the dairy herd are outlined in the diagram opposite (top). The distinctions between the three may be blurred, since some farmers with a suckler herd also rear purchased cattle, and some may have pedigree cows as well as a commercial suckler unit.

## The Beef Rearer

The beef rearer does not keep a breeding herd but buys in young cattle to rear, either to sell on for finishing elsewhere or to fatten and finish themselves.

The sources of cattle are:

- Dairy-cross beef calves from dairy herds. Although some dairy farmers rear these calves for a few weeks (often using a nurse cow or waste milk), most sell them, either at market or direct from the farm, at seven-plus days old. The beef rearer must therefore rear these calves to weaning (*see* Chapter 3). Some dairy herds milk 'dual-purpose' cows whose calves, when crossed

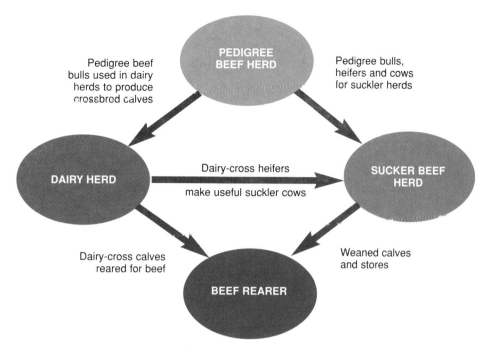

*Diagram showing the interactions between herd types in the UK beef industry.*

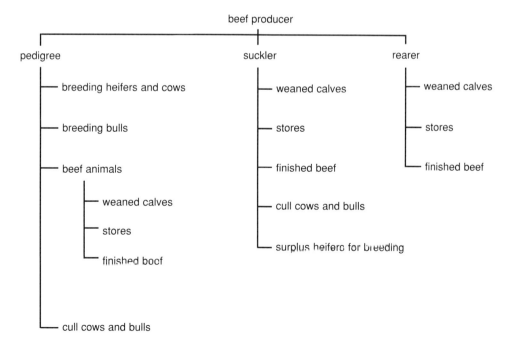

*The beef animals that can be sold from three types of beef herd.*

back to a beef bull, have good conformation and growth rates and are therefore well-suited to beef rearing.

- Suckler herds. Most farmers with spring-calving suckler cows sell their calves in autumn, at weaning, though some keep them on to heavier weights to sell them as stores for fattening. Calves from suckler herds tend to be better grown and with greater potential to finish as prime-quality beef than dairy-cross calves because they are usually not weaned until they are six to nine months old and are purebred or three-quarter-bred beef animals.
- Other beef rearers. Some animals are reared from calves to finishing on a single unit, but there is also a brisk trade of cattle between beef-rearing units, with individuals specializing, for example, in calf-rearing or in finishing strong store cattle.

Buying in cattle poses special risks of infectious disease. The beef rearer often buys from a variety of sources, so cattle meet new dis-

*Cattle fattening in a beef-rearing shed. These are crossbred Charolais, Simmental and Limousin steers, with large frames and a high growth-rate potential.*

*Hereford calf suckling. In the suckler herd, cows may single-suckle their own calves for six to nine months until weaning. Some farmers buy in extra calves so cows rear two or more calves at a time (double- or multiple-suckling).*

ease organisms to which they have no immunity. They are also likely to be highly susceptible to disease because they have been stressed by handling, transport and marketing and may not eat or drink for many hours. Suckler calves that have been extensively reared and never handled, and may have just been weaned, castrated and dehorned, may be particularly at risk of stress-related disease. Consequently management protocols designed to reduce disease in bought-in cattle should form an important part of any management protocol.

## The Suckler Herd

Most suckler herds calve seasonally, either in the spring or in the autumn. The nutritional demands of spring-calving cows are highest (at peak lactation) in spring/summer so they make best use of cheap, high-quality grazing. Calves are weaned in autumn and usually sold. The herd's total feed requirements are least during the winter (because there are no calves to be maintained and cows are dry), when feed is relatively more expensive (if conserved forage is fed), or poorer quality (if cows are overwintered at grass or on kale or turnips). Autumn-calving suckler herds wean

calves in spring when calf prices are high, and any check to calf growth following weaning is kept to a minimum because they are fed high-quality spring grass. Winter feed requirements are higher for the autumn-calving herd because cows are lactating rather than dry. Housing costs are also higher because cows are calving or have calves at foot, so their housing requirements are more demanding than for dry, spring-calving cows.

The economic goal for a suckler producer is to maximize the weight of calf produced annually by ensuring (a) that as many cows as possible calve each year, and (b) that optimal weight gain in the calf is achieved, and at the lowest possible feed cost.

### Feed Costs

Feed costs in suckler production account for 75 per cent of the variable costs and 50 per cent of total costs. Most of these feed costs are spent on maintaining the cow, not feeding the calf. To achieve low feed costs there are three fundamental principles to be applied:

1. The proportion of feed cost spent on maintaining the cow should be as low as possible.

*A tight calving pattern ensures that cows and calves can be managed and fed as a group – note that these calves are all of a very similar age. (The crossbred cows shown here were served by a beef bull to produce three-quarter beef calves.)*

It is very inefficient to have a large cow and produce small calves. Feed costs for maintenance depend upon breed: native breeds are usually small and adapted to thrive on poor-quality forage, so maintenance costs are low; however, their calves do not have the same growth potential as the calves of the large Continental breeds. For example, a 700kg cow requires 1.6 tonnes silage of average quality over the winter period compared to a tonne for a 500kg cow.

2. The maximum use of the cheapest feed grass should be made.

This is done by 'making hay in the good times'. When grass is plentiful it should be used to maximum effect, so that it not only supports the cow's current physiological requirement, such as maintenance plus lactation, but also allows the gaining of condition. Outside the optimal grazing season, when supplementation or housing is required, feed costs can be kept low by utilizing the fat reserves built up over the grazing period (i.e. allowing the cow to lose condition).

3. Cows should calve in a tight group so that they can be fed as one group to match their physiological requirements.

### Selective Breeding for Calf Conformation and Growth

This is discussed fully in Chapter 7, but issues that are important for the suckler herd are:

- Use of performance-recorded sires for consistent results.
- Choice of breed most appropriate for the farming system.
- Use of crossbred cows to capture hybrid vigour.

*(Modified from Pullar, 2002.)*

### Fertility

A tight calving pattern ensures that all calves are born within a short period, so that they are fairly even in size and age, and weaning weight can be maximized. They can be managed as a group, weaned at the same time, and sold in batches of even-sized animals (which tend to fetch better prices than mixed groups). The cows can also be managed as a group, for example given routine treatments (e.g. copper or magnesium boluses) and pregnancy-tested at the same time. Most importantly, the nutritional requirements will be broadly similar, since they are all at a similar stage of the production cycle, and can be fed as a group, thereby reducing feed costs.

In order to achieve a tight calving pattern, the service period, whether using a bull or

AI, must be short (generally six to nine weeks). Unless conception rates are high, this will leave many cows empty at the end of the service period. These cows must either be culled, which leads to high replacement costs, or must be kept empty for a year, incurring high costs to maintain unproductive animals. Beef cows that are dry for a prolonged period tend to get very fat, so even if they conceive the following season they are at high risk of calving difficulties (dystocia). It may then be tempting to extend the service period, but this leads to cows calving out of season (so the herd cannot be managed as a group) and a few much younger calves (whose individual requirements may not be adequately met).

Good fertility management is therefore essential. Fertility targets and the factors affecting fertility in the beef herd are discussed in detail in Chapter 7, but the most important are to ensure that bulls are fertile and that cow body condition is neither too fat nor too thin (not less than a body condition score (BCS) of 2.5 at service; condition is based on a 1–5 scoring system (*see* Chapter 5). Suckler cows at risk of losing excess weight in early lactation are older cows that cannot compete for winter feed, and first-calved heifers, particularly those of large Continental breeds, which will be growing throughout first lactation. These animals should be separated for extra feeding. The BCS of the whole herd should also be regularly monitored.

### Mortality Rate of Calves

Most suckler calves that die do so at calving, or in the first few days after birth, or as a result of an outbreak of infectious disease. Early calf disease and infectious disease are discussed in detail elsewhere (*see* Chapters 3 and 9), so only the causes and prevention of deaths during calving are explored here.

One cause of death is dystocia, when the cow fails to deliver the calf successfully. This occurs either because of a relative disproportion between the size of the calf and the size of its mother (in other words, the calf is too big for the cow/heifer) or because it is presented abnormally (*see* Chapter 7). Most cases of dystocia can be resolved, resulting in the birth of a live calf, provided that skilled assistance (from the farmer or the veterinarian) is given in good time.

In some cases, calves are born alive but then die because they do not take a breath, either because they are smothered by membranes that the mother fails to lick clear, or because they are weak.

Deaths at calving can be minimized as follows:

- Choose the right breed of cow. Certain breeds, particularly the hardy native breeds, tend to have small, vigorous calves that are keen to get up and suckle within minutes of birth. Cows of these breeds are usually good mothers, and lick the calf to stimulate breathing as soon as it is born. Though such calves have less potential for high growth rate than other breeds, they are particularly suitable for extensively managed herds or for part-time beef farmers whose cows may have to calve without supervision.
- Choose the right bull. The size of the calf is heavily influenced by the bull, so selection of an appropriate bull is paramount. A compromise must be made between a bull whose calves have a heavy birth weight, good conformation and high potential growth (and are therefore usually large with muscular shoulders and hindquarters), and one whose progeny are easy to calve but may not have such good conformation or potential for growth. Ideally a bull (or semen from a bull) for whom information on gestation length and ease of calving is available should be chosen.
- Keep accurate service records. A record of service date(s) allows the calving date to be predicted so that management changes can be made and the cow can be monitored. This is easy if AI is used but difficult with some bulls (who may never be observed to serve a cow). Use of a raddle (*see* page 125) allows prediction of calving date.

- Pregnancy diagnosis. This allows identification of empty cows and can predict approximate calving dates (provided it is undertaken at less than four months' gestation).
- Monitor condition score. Cows that are overfat are more likely to suffer dystocia because fat around the pelvic canal restricts the size of the birth canal, and fat mobilization makes them more prone to metabolic disease. Autumn-calving suckler cows are likely to put on weight because they wean their calves in spring/summer and are dry during the early autumn, when there is often a flush of grass. Cows that have a prolonged dry period (for example if they fail to conceive one season and are kept on for a year) are also at risk of excessive weight gain. Body condition should be monitored regularly throughout pregnancy so that feed can be restricted to make sure that the body condition score does not rise above 3.5 (*see* Chapter 5). Beef cows that become very thin or emaciated are likely to give birth to small and weak calves; they may be prone to metabolic disease; they may become exhausted during calving so fail to mother their calves adequately; and the quantity and quality of colostrum they produce is likely to be poor.
- Provide adequate trace element nutrition. Deficiencies of selenium/vitamin E and iodine are associated with birth of weak or stillborn calves. Unfortunately the problem is often not diagnosed until calf deaths have occurred (when it is found at postmortem examination or by blood-sampling dry cows), but can be prevented by supplying these micronutrients.
- Prepare calving accommodation. Calves are more likely to survive if they are born into a clean environment and if mothers are separated from other cows due to calve and/or have space to mother their calves immediately after birth. If calving at grass, the calving paddock should not be overcrowded and should provide some shelter. Facilities should be available to examine and assist a calving cow if necessary; many beef cows are not accustomed to being handled, so facilities should provide adequate restraint with minimum stress. A crush is not suitable for calving a cow because of the risk of injury if she goes down.
- Provide adequate supervision at calving. Cows should be monitored to ensure that signs that calving is imminent are observed so that calving can be supervised.
- Recognize the limits of your skills. Delivery of a live calf is very important in the beef herd. Once the calf is engaged within the pelvic canal there is only a limited time before it dies, so if you intervene but then find you cannot resolve a calving problem quickly, you should call for veterinary assistance sooner rather than later to maximize the chance of a live birth.
- Make sure the calf is breathing, gets up and receives enough colostrum. (*See also* Chapters 3 and 7.) First-calving heifers and cows that are exhausted after a prolonged calving may not adequately look after the calf once it is born. Other cows may 'steal' a newborn calf, so it may be necessary to separate the fresh-calved animal and her calf to enable them to bond.

Planning and attention to detail keep calf mortality to a minimum. If problems occur, advice should be sought.

Other factors that influence the efficiency of the suckler herd include:

- Culling rate of cows (which determines the number of replacements required).
- Replacement heifers.
- Culling rate of bulls.
- Replacement bulls.
- Mortality rate of cows.
- Infectious and parasitic diseases.
- Unwanted pregnancies in young heifers.

### Culling Rate of Cows

Cow culling rates can be very low (10–15 per cent) in suckler herds, with some cows successfully producing a calf every year until they are fifteen-plus years old. Poor fertility has already been discussed as a cause of forced culls, but other causes include lameness, wasting disease (such as Johne's), calving injuries, and chronic mastitis.

### Replacement Heifers

The number of replacement heifers required depends upon culling and mortality in the adult herd. Replacements may be home-bred or purchased from pedigree herds, from other suckler herds, or from dairy farms. The animals chosen should be of the best genetic potential and of the breed best suited to the farm's system of management; heifers should be reared according to the principles discussed in Chapter 4.

### Culling Rate of Bulls

The culling rate of bulls is often determined by policy for heifer replacements. If they are home-bred, breeding bulls will have to be replaced every two to three years to prevent inbreeding. If young bulls are purchased, they should still have a value to be sold on as breeding animals. Older bulls may become too heavy for use with heifers and must be culled. Despite their importance, bulls are often rather neglected in the suckler herd. They should be managed carefully to make sure they are fit for the service period and to minimize forced culls because they are expensive to replace. Lameness is common in bulls, particularly in some breeds. It must be treated early and thoroughly.

### Replacement Bulls

The replacement bull should be of the best quality possible because, as is commonly said, the bull is half the herd. (Bull selection is discussed below and in Chapter 7.) Some suckler herds utilize hire bulls, which avoids purchase and maintenance costs. However, hire bulls present a significant risk of brought-in diseases, including infectious causes of infertility, such as bovine viral diarrhoea (BVD) and infectious bovine rhinotracheitis (IBR).

### Mortality Rate of Cows

The target mortality rate for adult beef cows is less than 2 per cent. The major causes of fallen stock are deaths at calving, or cows that go down at calving and never get up ('downers'). This number can be minimized in the same way as deaths of calves are minimized during calving (*see* page 39). Hypomagnesaemia (low blood magnesium) is a common cause of sudden death in beef cows at pasture in autumn and early spring; it is often associated with lush pasture and wet stormy weather. It can be prevented by supplying magnesium in water. However, if cattle are able to access water supplies to which magnesium cannot be added (such as streams), it can be supplied in concentrate feed, or as a bolus. Magnesium licks may be effective but there is a risk that some animals may not use them.

### Infectious and Parasitic Diseases

Infectious and parasitic dizease should be controlled as discussed in Chapter 9.

### Unwanted Pregnancies

Unwanted pregnancies may occur in young heifers before weaning, where a bull running with the cows serves suckling heifer calves. This is a particular risk if the service period is protracted and with certain breeds (such as the Belgian Blue.) In some herds, all heifers are routinely treated to eliminate unwanted pregnancies at the end of the service period. Unwanted pregnancies may also occur where bull calves are not separated until late, or if the bull (or bulling heifer) escapes. It should be noted that no chemical method of removal of unwanted pregnancy is entirely reliable and heifers should be rechecked some weeks after treatment to make sure that they are empty.

## The Pedigree Herd

The primary aim of keeping a pedigree herd is to produce breeding stock for sale, with the emphasis being upon the quality of stock they produce, rather than the number, because top pedigree bulls achieve prices more than ten times that of commercial stock. Some farmers with pedigree herds add extra value by showing their animals either at agricultural shows or at bull sales, but the time and cost of preparing animals for such events is considerable and must be offset against potential extra income.

In general, the factors that influence the efficiency of pedigree beef production are exactly the same as those discussed for the suckler herd, with calf conformation and growth, fertility management and calf mortality being the key factors. However, there are some differences and these include:

• Breeding management.
• Record-keeping.
• Monitoring performance.
• Nutrition.
• Preparing animals for show and sale.

### Breeding Management

The pedigree herd may include several distinct genetic lines that are bred separately to produce animals with the desired pedigree. Season of breeding may be chosen to ensure that calves or young bulls are at the right age and weight for show or sale, rather than for cost-effective utilization of home-produced feed.

### Record-Keeping

Every animal must be recorded, and every potential breeding bull registered before he reaches an age specified by the breed society.

### Monitoring Performance

Although weight records and visual assessment by the skilled stockperson can give an indication of a bull's potential value as a breeding animal, analysis of the Estimated Breeding Value (EBV) allows a more accurate assessment of his potential. This is discussed further in Chapter 7.

### Nutrition

Rations are designed not only for cost effectiveness but to maximize rate of growth for a given age. Cattle must be regularly weighed, their

*Well-bred pedigree cattle, such as these Aberdeen Angus animals, can command high prices. This is an early-maturing breed, particularly suited to less-intensive systems. Aberdeen Angus bulls are often used to cross with dairy heifers because the bulls are not too heavy and their calves are not too big.*

condition assessed, and the ration adjusted as necessary to ensure that targets are met.

*Preparing Animals for Show and Sale*
This may take several months. Cattle are fed to achieve show condition (generally very fat). They must be halter-trained and taught to be led and handled by strangers. They must be housed in a clean environment so that their coats do not become soiled. They are groomed and shampooed and their hooves are oiled.

The farmer may sell pedigree bulls to dairy herds to produce crossbred dairy calves to be reared for beef, and bulls, cows and heifers to suckler herds and to other pedigree breeders.

# ECONOMICS OF BEEF CATTLE PRODUCTION

A partial budget of variable costs can be built up for beef herds in exactly the same way as it is for dairy cattle (*see* Chapter 1).

In simple terms, the net profit for a beef enterprise can be calculated as follows:

Net Profit = sale price of cattle − (purchase cost of cattle + total rearing cost)
Total rearing cost = cost per day × number of days.

Logically, therefore, profits should be highest if:

- Purchase cost of cattle (either the cost to buy or the cost to produce) is minimized.
- Rearing costs are minimized (either by reducing costs per day or by reducing the number of days for a given number of cattle).
- The sale price of cattle is maximized.

However, it is not so simple because these three factors are interlinked. For example, calves that are cheap to buy at market are likely to have a low genetic potential for growth, may be small or ill-thriven or diseased, and may not have been dehorned and castrated. Rearing costs are therefore likely to

be higher: small, unthrifty calves with low genetic potential grow slowly even if nutrition and husbandry are excellent. Veterinary costs may be incurred to treat disease carried by the calves. It is also costly to castrate and dehorn older cattle. Sale prices are also likely to be lower since such calves may never be transformed into top-quality beef animals. Thus, to buy calves at the bottom end of the market may be a false economy, although some farmers choose these animals, aiming to make a profit by keeping costs down while accepting a certain level of mortality and disease.

If the farmer chooses to buy or produce high-value calves, and invests in a high-cost rearing system, there should be a rapid turnover and cattle should command a high price when they are ready for sale. But prices for beef cattle are volatile (*see* graph on page 45), fluctuating at a particular market from week to week; if they drop just as cattle are ready, profit may be wiped out. If the farmer then keeps the cattle on in the hope that prices will improve, extra rearing costs will be incurred and cattle may go past their prime age and weight for sale.

Therefore it is more accurate to suggest that profits will be highest if:

- The best quality young cattle are bought or produced for a particular budget.
- Rearing costs are minimized (either by reducing costs per day or by reducing the number of days for a given number of cattle).
- The sale price of cattle is maximized.

Variation in cattle price is a major determinant of potential profitability of the beef herd, and its causes are explored in the following section.

## Factors Affecting the Price of Beef Cattle
There is an element of speculation and even luck involved in buying and selling beef cattle. Fluctuations in price may allow for profits if calves were bought cheap and are ready for sale when prices are high, but may result in

*Where horned cattle are mixed with animals without horns, stocking density must be low in order to reduce the risk of injury. In the UK, only a veterinarian is permitted to dehorn older cattle, so it is a costly procedure.*

losses if calves were expensive to purchase and prices have dropped when they are ready for sale. Usually when the price for finished beef is high, the price of calves, young beef animals and stores is also high. The margin for profit is often small, particularly if the farmer aims to keep cattle for only a few weeks before selling them on.

The causes of variation in the price of beef cattle are illustrated in the diagram on page 28 (top). The individual farmer has little or no influence on some, such as the world market price for beef, supermarket policy for beef sales, or national political climate. In the UK, the opening of the export market for calves in 2006 and falling numbers of dairy cows have led to a reduction in the availability of dairy-cross beef calves; as supply falls, prices are likely to rise. As prices rise, farmers whose main enterprise may be sheep, dairy or arable, may see rearing beef cattle as a poten-

tially lucrative diversification. It is relatively easy to dip into beef farming, utilizing buildings that are generally used for other purposes. This may further alter demand and supply, and influence beef prices. There may also be unpredictable seasonal effects. For example, farmers may be forced to sell cattle if spring arrives unexpectedly late and winter forage is in short supply, with the result that prices drop. In a good grass-growing summer/autumn there may be a surge in demand (and consequently in price) from farmers who want to make best use of extra pasture.

Other factors that influence beef cattle prices include:

- Breed. Pedigree animals sold for breeding are likely to achieve the highest prices. Pure-bred cattle sold as beef animals (for example, Aberdeen Angus or Hereford) may be eligible for niche markets and consequently

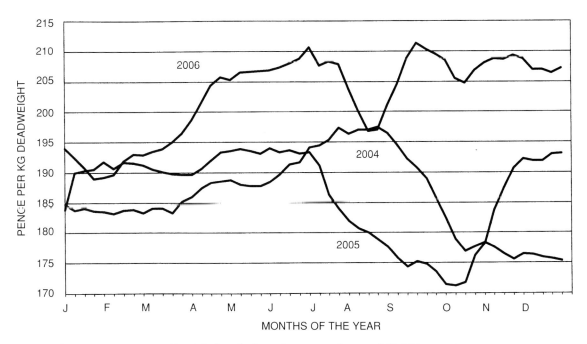

*Graph showing variation in British finished-cattle prices. (Source: MLC Economics, courtesy Mark Topliff.)*

achieve high prices. Continental breeds (either dairy crosses or purebred) that have a high potential for growth tend to command high prices. Holstein-cross cattle may be difficult to finish, so may have a lower value than other dairy-cross animals.

- Sex. The highest prices are achieved by pedigree breeding bulls that have won at bull shows or come from herds that have a record of winning shows. Other cattle sold for breeding may or may not command higher prices than beef animals, depending on current market conditions. For beef cattle, the price paid for heifers tends to be lower than for steers because they are smaller and finish at lower weights. Bull beef production is a specialist enterprise, so prices for dairy-cross bulls may also be lower.
- Age and weight. Older, heavier beef animals should always command a higher price than younger, smaller ones, though market fluctuations mean that this is not

always the case. Cattle that are well grown for their age achieve better prices than those that are poorly grown. The price paid for cull animals depends upon weight and fat cover. In the UK, aged animals (born before August 1996) cannot be sold for meat, as a precaution against BSE.

- Disease status. Premium prices may be achieved for breeding animals from herds of known disease status: there are health schemes that monitor cattle for presence of diseases such as Johne's, BVD, IBR and leptospirosis and require farmers to adhere to a strict biosecurity code. Farmers may choose to pay extra (compared to market price) to buy calves directly from an individual herd that they know, confident in the quality of husbandry, management and disease status. Cattle from TB-infected herds can be sold only under licence to herds that rear them under prescribed conditions, so prices are likely to be lower. It is illegal to transport or sell at market cattle showing signs of disease. Animals that may have previously been sick (and appear at market in poor body condition with dull coat, or crusts of dried dung around the tail indicating recent diarrhoea) are, of course, unlikely to make good prices.

- Rearing system. Organic or 'extensively reared' cattle generally achieve higher prices than conventionally reared cattle. Consumers increasingly care about animal welfare and will pay extra for produce from welfare-friendly systems. Farm assurance schemes demand a set of health, welfare and food safety standards for livestock production. Herds that comply with these standards may achieve premium prices or may have access to markets from which other herds are excluded.

- Direct sales. Farmers who sell direct to the public cut out the retailer's share of profits and are likely to achieve premium prices. There has been an escalation in the prevalence of farmers' markets and farm shops. Speciality cuts and rare-breed stock often sell at high prices; but the extra revenue must be offset against the time involved in advertising, marketing and selling.

### Factors Affecting the Cost of Home-Produced Calves

Pedigree and suckler herds produce their own calves, either to be sold or to be finished on-farm. The cost of these calves includes management and maintenance of the adult herd as well as the rearing or buying of replacement breeding animals. These herds do not incur the cost of buying in youngstock (as beef-rearing units do), but their extra costs include:

- Feed. There is the cost of feeding cows to supply nutritional requirements for maintenance, pregnancy and lactation, and the cost of feeding the bull and any replacement breeding heifers and bulls that are reared. Season of calving has a major influence on feed costs for the adult herd. Pedigree herds may be unable to make use of cheap sum-

mer grazing if they plan to show youngstock during the summer months.

- Replacements. Replacement breeding cows or heifers may be purchased or home-bred. The advantage of purchasing animals is that new genetic material is introduced. Many suckler herds purchase crossbred heifers from dairy herds: the price is generally reasonable. The cost of the home-bred replacement includes feed costs for her mother and rearing costs until she is ready to join the adult herd. In herds where AI is not used, replacement bulls are also required. The number of replacement animals depends upon fertility and longevity of adult cows. Replacement bulls are usually purchased rather than home-bred – to avoid inbreeding – but they can be home-bred using AI on a cow that has no familial connection with most of the female herd.

- Veterinary treatment. Most veterinary costs for the adult herd are incurred around calving time. Calving problems are most likely with breeds that produce large calves, particularly those with double-muscling, but these calves have the potential to grow fast and achieve high prices.

- Bedding. Hardy beef cattle in spring-calving herds may be outwintered or require only basic shelter, so costs vary widely according to season of calving, breed and local climate.

The most important determinants of cost to produce beef calves are the fertility of the adult herd and the number of calves that are born dead, or die. Targets of 95 per cent calves born per 100 females mated, and 3–5 per cent calf mortality, can be achieved and will keep the cost of producing home-bred calves to a minimum. Factors that may influence fertility rate and calf mortality are explored above and in chapters 3 and 7.

### Factors Affecting Rearing Costs

At the beginning of this section (*see* page 43), rearing costs were broken down as costs per day multiplied by the number of days. The

*Direct sales at farm shops and farmers' markets are increasing. Consumers are willing to pay higher prices for produce whose provenance is known, and direct sales eliminate the 'middle man'.*

costs per day can be built up much as for the home-produced calf:

- Feed. This may be purchased or home pro-duced. The cost of purchased feed is the price paid plus any transport and handling fee. The cost of home-produced feed includes seed, fertilizer and pesticides, labour and fuel for cultivation, harvesting, and so on. In herds that purchase cattle, feed costs include artificial milk replacer if calves are pur-chased before weaning, and forage, concen-trate or 'straights' to supply requirements for maintenance and growth of beef cattle. The level and quality of feed supplied will af-fect its cost but also affect the growth rates that can be achieved.
- Veterinary treatment. This includes both preventive measures (such *as vaccines and wormers) and treatment of sick animals. Health planning and biosecurity measures (*see* Chapter 9) can help to minimize veteri-nary costs.
- Bedding and incidentals. As for home-produced calves, these depend upon the rearing system, with extensively reared cattle utilizing little or no bedding.
- Transport and marketing. These may be significant for beef cattle rearers who buy

and sell large numbers of cattle or where turnover of stock is rapid.

The number of days required to rear cattle to a given weight or finish is determined by growth rate. This in turn is determined by a series of factors including breed, sex, age, nu-trition, disease and trace element status.

## Veal Production

In the UK there is a growing market for veal, mainly for gourmet butchers and the restau-rant trade. In the UK, veal calves are group-reared on straw with a milk-based diet that includes some fibre and must include at least 40g iron per day. Calves are slaughtered at sixteen to twenty-two weeks. Some organic veal is also produced from suckled calves.

The market for beef is robust, provided that it meets the specifications required by abat-toirs, retailers and, ultimately, the consumer. Producers can maximize potential profits by selectively breeding or purchasing the best-quality stock, by providing excellent nutrition, and by controlling and preventing disease.

Realistic performance targets should be set individually for each unit, taking account of factors such as grazing and forage quality,

*Growing beef cattle, confined to the yard while their feed troughs are filled.*

access to cheap straw, housing, and skill and availability of labour. The targets will, of course, depend upon the classes of stock kept but will include measures of:

- Fertility.
- Calving problems.
- Mortality rate of cows around calving.
- Culling rate.
- Growth rate.
- Carcass fatness and conformation.

- Mortality rate of calves and its causes.
- Other mortality and its causes (where known).
- Incidence of diseases – such as lameness, pneumonia and scour – that have a profound effect on production and profits.

Progress should be monitored and recorded to check that targets are being met; if not, the causes of problems should be investigated and action taken to correct them.

# CHAPTER 3

# The Calf

The causes of many calf problems are simple compared, say, with the complex issues that may cause disappointing milk yields from dairy cows, yet calf problems are common. In this chapter, management of the calf from birth to weaning is examined.

## THE NEWBORN CALF (0–24 HOURS)

The newly born calf takes its first gasping breath, and within five minutes it is breathing regularly, with its head raised. It may start to vocalize and, often, within twenty minutes it struggles to its feet, the mother continuing to clean it; by the time it is thirty minutes old, the calf may be up and actively searching for the teat. A vigorous calf will be suckling by the time it is forty-five minutes old.

It is best to allow cow and calf to bond without interference. However, there are times when it is essential to intervene, for example following a difficult calving where cow and calf are exhausted. Breathing can be stimulated by clearing the nose and mouth of mucus and rubbing the calf's chest with straw. If its nose is tickled with a piece of straw, or cold water poured into its ear, the calf will usually take a first breath. The airways may be clogged with fluid following a prolonged calving, and this can be cleared by hanging the calf upside down over its mother (if she is still down) or over a gate (great care must be taken to prevent injury that can occur if a calf is flung roughly over a gate). Sometimes a remarkable volume of fluid drains out – if left, the calf will

not be able to inflate its lungs fully and may die or develop pneumonia. Once breathing, the calf must be dried by rubbing it with straw because the newborn calf rapidly gets cold if it is left wet with birth fluids. Usually the calf then stands and suckles in the normal way. However, if the cow refuses to allow the calf to suckle, or the calf is too weak to feed, it must be fed with colostrum (*see* page 51).

### Calving Environment

The calf should be born onto a clean dry bed to minimize the risk of infection entering via the

*Calf aged five minutes. Its head is already raised, and within a short time it will be trying to stand.*

navel or at suckling soon after birth, so calving boxes should be cleaned out after every three to four calvings (at least), and the cow (particularly her teats) should be clean. (N.B. a deep-litter bed or sand bed covered with deep clean straw is ideal for the calving cow, who needs a non-slip surface in the calving box.) If housed, the cow should be separated from other heavily pregnant cows so that they do not try to steal the calf and prevent it from suckling. Easy access to facilities for restraint of the cow is essential. A calving paddock may not always provide a clean environment if it gets poached and muddy or is overstocked.

*Dead calves in a dairy herd. Calves may die at birth or they may perish later from diseases such as diarrhoea or navel ill. The costs of mortality are high, and its causes should be investigated and tackled.*

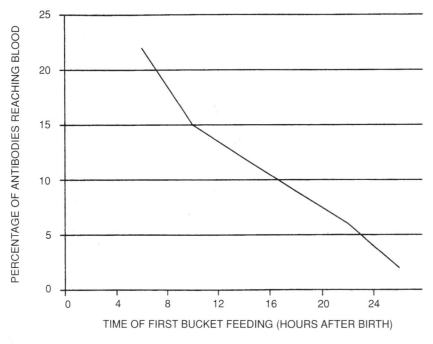

*Graph showing the effect of first feeding time on uptake of antibodies.*

Hold the calf steady with body to leave both hands free to help it suckle.

Guide the calf's mouth to the teat and squirt in colostrum. A vigorous calf learns quickly.

*Helping the calf to her first food. This heifer is keen to mother her calf: she is quiet and needs no restraint, but neither she nor her calf knows what to do at first. (Photo: Ulrike Wood.)*

## Colostrum

The calf is born with no immunity to disease. Immunity is derived solely from intake of colostrum (first milk) during the first twenty-four hours of life. The newborn calf's intestine can absorb antibodies directly into the bloodstream. This property of the gut wall is progressively lost over the first twenty-four hours of life. After twenty-four hours only smaller molecules, broken down by digestion, can be absorbed. It is therefore essential that the calf receives plenty of colostrum within the first six hours of birth, when absorption of antibodies is maximal. Although no further antibody absorption occurs, it is valuable to continue to feed colostrum for two to three days, whose antibodies provide local protection for the lining of the gut, preventing invasion by bacteria.

Following a normal calving, a vigorous calf is up and suckling strongly soon after birth. However, it may need help if:

- The calf is unable to find the teat. This is a particular problem with older cows whose udders are pendulous and large.
- The cow (especially first-calf heifer) will not stand to be suckled and may need to be restrained.
- The cow is down, owing to exhaustion or calving injury.
- The calf is weak or exhausted.

If there is any doubt about whether a calf has received enough colostrum, it should be fed either with a bottle or by stomach tube (*see* below). Antibody absorption is better by bottle, so this method should be used if the calf is able to suck. A supply of frozen or artificial colostrum should be kept for emergency use. (Frozen colostrum should be thawed in warm water, but must not be overheated as excess heat destroys the antibodies.)

*How to Use a Stomach Tube and Bag*

1. Fill with 1.5 litres of colostrum and close clip.

2. Dip the tube in hot water to make it flexible.

3. Hold the calf steady between your legs and introduce the bulb of the tube into the side of the mouth and into the oesophagus (gullet).

| | Table 4: Target colostrum intake in first six hours | |
|---|---|---|
| Calf weight | Colostrum intake in first 6 hours | Volume per feed |
| 30kg | 3 litres | 1.5 litres |
| 40kg | 4 litres | 2 litres |
| 50kg | 5 litres | 2.5 litres |

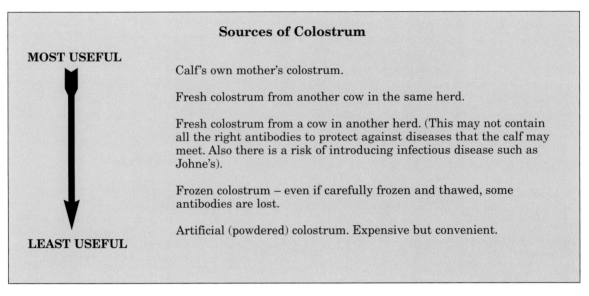

**Sources of Colostrum**

MOST USEFUL

Calf's own mother's colostrum.

Fresh colostrum from another cow in the same herd.

Fresh colostrum from a cow in another herd. (This may not contain all the right antibodies to protect against diseases that the calf may meet. Also there is a risk of introducing infectious disease such as Johne's).

Frozen colostrum – even if carefully frozen and thawed, some antibodies are lost.

Artificial (powdered) colostrum. Expensive but convenient.

LEAST USEFUL

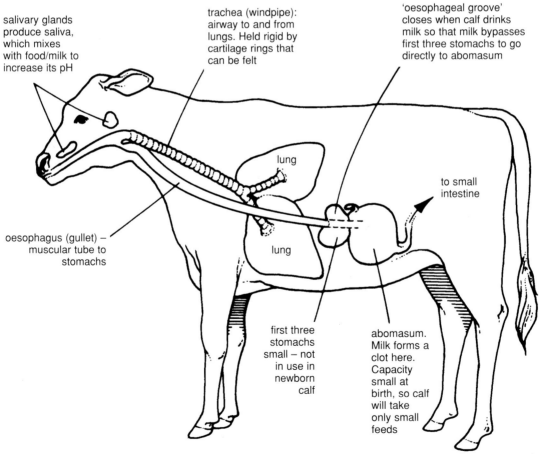

salivary glands produce saliva, which mixes with food/milk to increase its pH

trachea (windpipe): airway to and from lungs. Held rigid by cartilage rings that can be felt

'oesophageal groove' closes when calf drinks milk so that milk bypasses first three stomachs to go directly to abomasum

lung

to small intestine

oesophagus (gullet) – muscular tube to stomachs

lung

first three stomachs small – not in use in newborn calf

abomasum. Milk forms a clot here. Capacity small at birth, so calf will take only small feeds

*The anatomy of the young calf.*

4. With one hand around the calf's neck, check that the stomach tube is not in the trachea (windpipe). The trachea runs parallel to the oesophagus and is kept open by rings of cartilage. If the stomach tube is in the trachea, you cannot feel it through the cartilage rings. If you can feel the end of the tube, it is correctly placed in the oesophagus. (Don't rely on the calf to let you know – it may not struggle or cough if the stomach tube is in the trachea, and if you run colostrum through the trachea into its lungs, it will drown.)

5. Continue to introduce the stomach tube into the calf to about 15–20cm.

6. Open the clip to allow colostrum to run. If it fails to flow, just move the tube up and down a little and it will usually start.

7. Once the bag is empty, close off the clip before withdrawing the tube.

*Colostrum Quality*

Even if the calf receives colostrum in the crucial first few hours after birth, it may not receive adequate antibody protection if the quality of the colostrum is poor. The antibody content of colostrum may vary from 20–100g per litre. Poor colostrum quality may occur if:

- The cow is in very poor body condition (*see* Chapter 5) at calving, particularly if she is short of vitamin E or selenium.
- The cow or heifer has been pre-milked, or colostrum has been leaking prior to calving. First-milking colostrum contains the highest concentration of antibodies, but also contains unique proteins that protect the antibodies from being digested in the calf's intestine, allowing a higher proportion to be absorbed intact.
- The cow or heifer was purchased recently before calving, so her colostrum may not contain specific antibodies against diseases on this particular farm.

- The animal is a heifer rather than a cow: antibody protection in heifers is generally less than it is in the cow. This is because the older cow has met more infectious agents and as a result produced more antibodies.
- The cow is a very high-yielding animal, whose colostrum is dilute, with a low concentration of antibodies.

Dirty environmental conditions may also reduce the effectiveness with which the calf's intestine takes up antibodies. The calf's intestine is sterile at birth, but is quickly colonized by microbes – the number of which depends on the hygiene of the environment. There is some evidence that bacteria in the gut hasten the rate at which the gut wall closes to prevent antibody absorption.

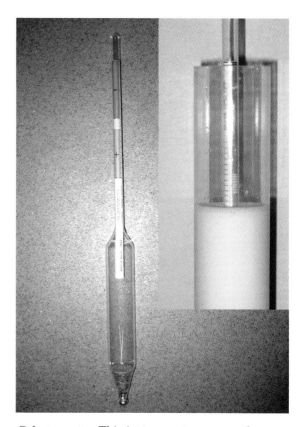

*Colostrometer. This instrument measures the specific gravity of colostrum to check the level of antibodies: green is good to adequate; yellow is suspect; red is poor.*

## Vaccination

Additional antibody protection may be provided by vaccinating the mother so that her colostrum will contain antibodies to protect the calf against specific infectious diseases.

## Navel Dip

The calf's navel is wet at birth, and opens directly into the calf's abdomen, providing a route for bacteria to invade. As soon as possible after birth, dip the navel with a strong solution of iodine to disinfect it, and then dry it. (*Do not* use teat dip, which contains emollient and prevents drying.) The alternative to dipping is to use a navel clip. Hygiene in the calving box is also important to minimize the bacterial challenge to the navel.

# FEEDING

## Milk

The suckler calf receives all its nutrition for the first weeks of life through its mother's milk. The calf suckles whenever it is hungry, (usually about six times every twenty-four hours), milk is at body temperature, freshly produced and (from the healthy cow) more or less sterile. The process of suckling closes the 'oesophageal groove', which effectively shuts off the first three stomachs so that milk is delivered directly into the fourth stomach, the 'true' stomach or abomasum (*see* diagram, page 52). The milk is mixed with saliva during suckling, increasing its pH. Renin is an enzyme produced in the wall of the abomasum and causes the milk to clot. The clot is then gradually broken down over the next two to four hours and passes into the small intestine a little at a time so that digestive function is not overloaded.

Artificial rearing systems for calves should mimic this natural process. Ideally therefore, artificial feeding should give:

- Frequent feeds.
- Evenly delivered feeds throughout the twenty-four hours.
- A system that closes the oesophageal groove.
- A system that produces a good clot in the abomasum.
- Milk at consistent temperature.
- Milk of consistent quality and concentration.
- High-quality protein for good growth.
- Fat globules of the correct size for efficient digestion.
- The right quantity each feed to prevent gorging or inadequate nutrition.
- Excellent hygiene.

However, the artificial feeding system must be as cost-effective and easily managed as possible, so compromises must be made that nevertheless produce healthy, well-grown calves. A number of feeding systems that are commonly used are evaluated in Table 5 (*see* page 56).

## Removing the Calf from the Cow

The dairy calf must be taken from its mother so that the cow can join the milking herd. Although her milk is not legally fit for the bulk tank until four days after calving, the calf is often removed before that time, but should stay on the cow for at least twenty-four hours. In some large herds the calf is commonly removed earlier than this, but colostrum must be fed for a minimum of twenty-four hours. It requires time and patience to get the calf drinking from a bucket.

Fresh water should always be available to young calves.

## Risks During the Milk-Fed Period

Mortality and disease rates are highest in young calves before weaning, and many of the problems are caused by, or exacerbated by, incorrect feeding practice. Mostly this is owing to inadequate attention to detail. For example:

- Incorrect mixing of milk powder. Milk substitute that is made up too dilute, too concentrated or incompletely mixed may cause digestive upset. Powder should be weighed every time. Using a scoop may

*Acidified milk replacer fed ad lib via multiple teats. One teat is provided for each calf so that they can all feed at once. Acidified milk replacer is fed once or twice daily.*

lead to inaccurate mixing because powder at the bottom of the sack is compacted, so a scoop from the top of the sack contains less powder than a scoop from the bottom.

- Underfeeding. Calves that are underfed may fail to thrive and be more prone to diseases such as pneumonia.
- Overfeeding causes milk to overload into the small intestine, causing diarrhoea.
- Changes in feed (for example between whole milk and milk substitute or between different types of milk substitute) may cause diarrhoea. Gut enzymes require several days to adjust to new feed so any change must be made gradually.
- Inadequate hygiene. All feeding implements for bucket-reared calves must be thoroughly cleaned twice daily after feeding. Each calf should have its own bucket. Particular care should be taken when dealing with any calf that has diarrhoea.

## Introduction of Solid Feed

Initially the calf receives all its nutrition from colostrum and then milk. The calf starts to nibble forage – grass, hay or straw – within a few days after birth. If concentrate is offered, the calf will begin to eat a small amount from seven to ten days old. Unlike milk, which passes directly to the abomasum, solid food passes into the rumen (first stomach). The fibre in the forage (so-called 'scratch factor') and the products of digestion promote development of the rumen's structure and function, and over the following weeks it becomes a vat for microbial fermentation (as in the adult digestive tract – *see* Chapter 5).

The suckler calf continues to receive most of its nutrition from milk for some months, but the dairy replacement heifer or other artificially reared calf is generally weaned at six to ten weeks old and it is important to stimulate early development of rumen function.

## Table 5: Artificial feeding systems for calves

| Milk type: | Key points | Advantages (+) and disadvantages (−) |
|---|---|---|
| Fresh waste whole milk | • Bi-product – often plenty available<br>• Variable quality – colostrum, antibiotic milk<br>• Dairy cows that have a chronic problem (lameness, mastitis) sometimes used to suckle calves twice daily | + Cheap<br>− Possible transmission of Johne's disease<br>− Variable quality may lead to digestive upset<br>− Variable quantity available<br>− Antibiotic milk may lead to antibiotic resistance<br>− Certain mastitis organisms may be transmitted to heifers through milk feeding |
| Stored waste whole milk | • Forms sour milk (*see* yoghurt mix below)<br>• Mix container daily<br>• Feed ad lib<br>• Thoroughly clean and start again every three to four weeks | + Cheap<br>− Some variation in quality<br>− Intake may fall between cleaning (becomes more acid and less palatable) |
| Yoghurt milk | • Whole milk or colostrum with added live yoghurt culture.<br>• Careful management to ensure that culture remains active<br>• Becomes slightly sour and thick<br>• Never allow the drum to empty, (retain about 10 per cent) add extra milk<br>• Mix daily<br>• Don't add antibiotic milk (kills culture)<br>• Thoroughly clean and start a fresh culture every four weeks | + May reduce incidence of scouring<br>− Intake may fall when mix has been active for several weeks (becomes more acid and less palatable) |
| Milk replacer | • Must be made up exactly to maker's specifications or digestive upsets<br>• Hygiene of all utensils and attention to detail paramount<br>• Store dry<br>• Many products available – different composition and different prices<br>• Skim-milk based powder forms clot<br>• Non-skim powders (containing whey-based and other protein sources) do not form a clot. Quality of protein and fat determines ability of calf to digest effectively | + Milk that forms a clot is digested more slowly with less risk of digestive upset<br>− Risk of overloading small intestine when non-skim milk replacer fed to calves less than four weeks old<br>− Expensive |
| Acidified milk replacer | • Must be made up exactly to maker's specifications or digestive upsets<br>• Hygiene of all utensils paramount<br>• Store dry | + Allows cold feeding once a day<br>+ May reduce incidence of scour<br>− Expensive |

| | Key points | Advantages (+) and disadvantages (–) |
|---|---|---|
| **Frequency:** | | |
| Once a day | • Once daily feeding not adequate | + Cheap on labour<br>– Often insufficient milk fed to meet growth potential<br>– Single large feed likely to cause digestive upset |
| Twice a day | • Commonly practised<br>• Ideally feed every twelve hours | + Enables you to check calves twice daily and monitor appetite, signs of disease, etc<br>– Takes time |
| Continuous | • Supply tank with teats<br>• Calves help themselves at will | + Labour-saving<br>+ Calves can take larger volumes so growth rates good<br>– May be difficult to maintain hygiene – easy spread of infectious organisms<br>– Impossible to monitor or regulate milk intake (some calves may gorge, others may not get enough) |
| Automatic feeder | • Computerized system<br>• Computer recognizes calf collars and allows calf to feed every two to four hours, delivering a fixed quantity of freshly mixed milk via a teat | + Closest to natural suckling<br>+ Can monitor and regulate intake for individual calves<br>+ Labour-saving<br>– Expensive to install |
| **Temperature:** | | |
| Cold | • Suitable for certain artificial milk powders | + Less growth of infectious organisms<br>– Not ideal for very young calves |
| Warm | • Blood heat | + Best for calves provided temperature is always the same |
| Variable | • Whole waste milk may sometimes be fed warm, sometimes cold | – Not ideal |
| **Quantity per feed:** | | |
| 2 litres increasing to 4 litres for older calves | • Maximum per feed for young dairy calf | – Must be fed two-plus times daily to provide adequate nutrition<br>– Restricts growth rate at an age when food conversion efficiency is very high |
| Ad lib | • Continuous feeding systems | – Can achieve high growth rates, which may have long-term benefits<br>– Possible gorging and digestive upsets when calves are young |

*The calf starts to eat forage from a few days old, stimulating rumen development and supplementing nutrition. Milk continues to supply most of the suckler calf's nutritional needs for several months, but the artificially reared calf must be ready to wean at six to eight weeks old.*

Fresh hay or barley straw should be offered to provide forage, and a small amount of fresh concentrate fed twice a day. Bucket-fed calves can be offered concentrate after each milk feed. Proprietary calf feeds are highly palatable and easily digested because energy and protein sources are based on milk proteins and lactose, and are designed to stimulate rumen development. Protein level of these feeds is 17–18 per cent, and they are fed either as a coarse mix or as compounded calf pencils. Although expensive, feed conversion of the young calf is very efficient and it is important to give these animals a good start. Feed must not be contaminated by dung, and stale feed must be regularly replaced.

The calf's diet over the next weeks and months depends upon its future purpose and the system of management, but gradually the proportion of the calf's requirements supplied by milk falls, while forage, with or without concentrate, increases.

## Weaning

The suckler calf is weaned from its mother at up to nine to ten months old to allow the cow a dry period of two months or more before the next calf is born. Although the calf is well able to thrive without milk, it will not want to be parted from its mother, so it should be weaned into a secure paddock or building in order to minimize the risk of escape and/or injury.

Because artificial rearing systems for calves are labour-intensive and costly, artificially reared calves are weaned from about six weeks old – much earlier than suckler calves. The calf should not be weaned until it is consuming a minimum of 1.25kg concentrate per day for three consecutive days (although many wean when calves reach an intake of 1kg daily) and has been observed regularly chewing the cud. While some wean the calf abruptly, others advocate reducing milk intake over several days, either by reducing the number of feeds or, with computerized feeders, by offering more dilute milk replacer. In either case, if the calf is weaned before its rumen is fully developed, it may become pot-bellied and unthrifty, and may scour (the so-called post-weaning scour syndrome).

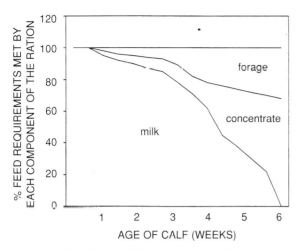

*Feed intake of the artificially reared calf, weaned at six weeks.*

## HOUSING THE CALF

For the first twenty-four hours after birth the calf is with its mother and requires only a dry bed and excellent hygiene. A calf born outside does not even need a dry bed, provided it can snuggle up to its mother and shelter from wind and rain.

There are many options for housing the artificially reared calf once separated from its mother, but the key elements to success are a dry, draught-free bed and good ventilation and drainage. An isolation pen is essential because the sick calf may shed millions of infectious organisms, for example via diarrhoea or coughing, and should be kept separate to prevent spread of disease to others. Ideally the calf should have separate air space and drainage from the rest of the group.

## ROUTINE PROCEDURES

### Castration

Some farmers rear pedigree bulls for sale, and some rear entire males as bull beef, but all other male calves must be castrated for ease of management: to prevent unwanted pregnancies and to avoid the development of aggressive bull behaviour. Calves should be castrated at as young an age as possible, when it is a relatively minor job causing little setback to calf growth and with few risks of complication. The calf should be adequately restrained in a calf crush or in the corner of a pen, against a

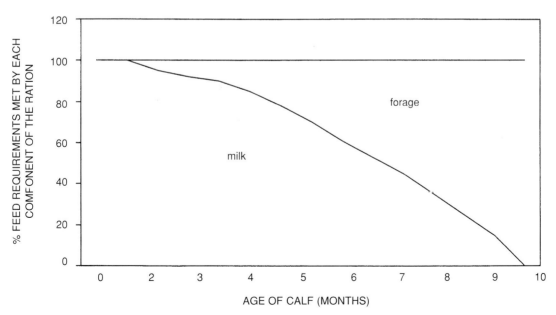

*Feed intake of the suckler calf. Concentrates are not essential, although some farmers offer calves a small amount via a 'creep' system, which gives access to the calf but not to the cow.*

*Concentrate feed for calves. A small amount of fresh, high-quality concentrate feed should be provided twice daily to encourage maximum intake.*

wall, so that it does not struggle and the castration can be carried out quickly. The veterinarian may use light sedation for large calves to aid handling. Calves should not be castrated immediately before or after weaning, since it creates additional stress.

Castration is easiest using an elastrator and rubber ring when the calf is less than seven days old. Both testicles must be below the ring and the urethra (which carries urine from the

*Calf hutches provide a dry, draught-free bed, with plenty of ventilation, and they are easily cleaned.*

bladder to the penis) must not be caught by the ring. Although there is no legal obligation to use local anaesthetic for castration of calves under a week, it is less stressful if local anaesthetic is injected into two to four sites at the level of the rubber ring (*see* diagram opposite).

Alternatively, the calf can be castrated using the Burdizzo (so-called bloodless castrator) or by surgical removal of the testicles. In either case, the calf must be adequately restrained and local anaesthetic must be used. The Burdizzo works by crushing the nerve and blood supply to the testicles, which causes them to gradually shrivel and disappear. Risks with Burdizzo castration are painful swelling of the scrotum, especially in older calves, and incomplete castration, usually because the cord has escaped the Burdizzo or because the cord has not been crushed for long enough. Risks of surgical castration are infection and bleeding, but these are minimal if the calf is adequately restrained, and if hygiene is good.

### Welfare Aspects of Castration

The calf should be as young as possible and well restrained to prevent struggling and to reduce stress. Local anesthetic should be used even in young calves less than a week old because it is a painful procedure and there is no

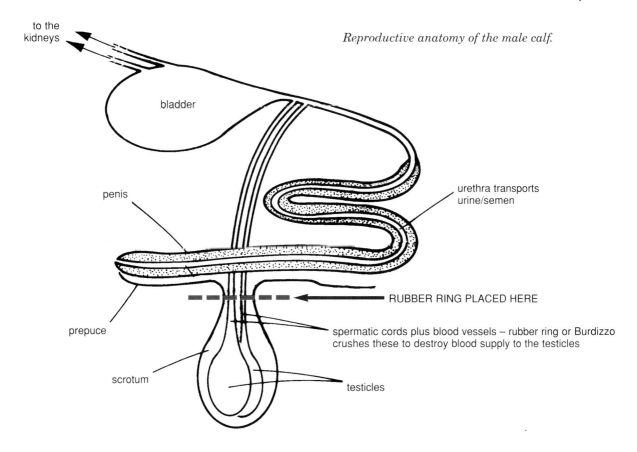

*Reproductive anatomy of the male calf.*

to the kidneys

bladder

penis

urethra transports urine/semen

prepuce

RUBBER RING PLACED HERE

spermatic cords plus blood vessels – rubber ring or Burdizzo crushes these to destroy blood supply to the testicles

scrotum

testicles

evidence to indicate that young calves feel less pain than older ones. Local anesthetic reduces pain sensation for only a short time: dogs, cats and horses are generally given pain relief after castration, though for economic reasons this is not usually offered to calves.

## Disbudding

Cattle are easiest to manage when they are without horns; there is less risk of injury to humans and other cattle, and space requirements are marginally less. There are naturally polled breeds (such as Aberdeen Angus, some Herefords and British White), but in most other cattle the horns are removed. This is best done with a hot iron when the horn buds are small but clearly defined, usually at four to six weeks old. It is a skilled job to burn the horn bud sufficiently to prevent regrowth of horn, but without damaging the calf's brain or skull by over-vigorous disbudding.

*Disbudding Method*

1. Restrain the calf well using either a calf crush or a halter to hold the head still.

2. Inject local anaesthetic into the space between the outer margin of the eye and the ear (*see* diagram on page 62), drawing back on the syringe to make sure the needle is not in a vein (the vein runs close to the nerve and intravenous injection of local anaesthetic is harmful).

3. If the calf has a very hairy poll, you may need to cut/clip the hair away to prevent it from catching fire.

4. Start heating the disbudding iron in a safe place where there is no risk of fire.

5. Wait ten to fifteen minutes to allow the local anaesthetic to work.

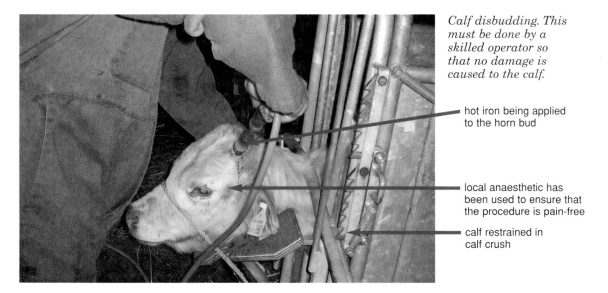

*Calf disbudding. This must be done by a skilled operator so that no damage is caused to the calf.*

hot iron being applied to the horn bud

local anaesthetic has been used to ensure that the procedure is pain-free

calf restrained in calf crush

6. Check that the disbudding iron is red hot and that local anaesthesia is effective. (If the anaesthetic is effective, the calf will barely struggle.)

7. Place the disbudder on the horn bud and burn the margins of the bud using a circular motion. This usually takes about fifteen seconds.

8. Use the disbudder to flip out the horn bud.

9. Once both sides are disbudded, apply antibiotic spray to reduce risk of infection at the sites.

10. Allow the calf to rest quietly to recover.

## Dehorning

It is possible to remove the horns from bigger calves and even adult cattle, but a veterinarian must perform these jobs. Dehorning large cattle may cause a setback in the animal's growth, and in adult cattle there is a risk of bleeding and of infection in the sinuses of the head. Clearly there is a risk to anyone handling large cattle for dehorning.

## Removal of Supernumerary Teats

The replacement heifer should have four well-placed teats, but some heifer calves are born with one, two or more extra teats. These are usually removed because:

✗ Site for injection of local anaesthetic, midway between the ear and the outer margin of the eye

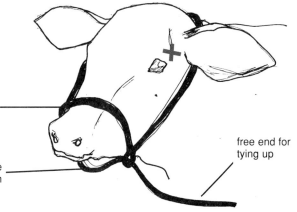

fixed section over calf's nose

free end for tying up

section next to free end under calf's chin

*A correctly fitted halter.*

1. They may interfere with placing of the milking cluster in dairy animals, and

2. They may be attached to mammary tissue that can develop mastitis.

The skin around the extra teats should be infiltrated with local anaesthetic and the extra teats snipped off using sharp scissors. It is important to be quite sure that the main teats are not removed – if there is any doubt, the procedure should not be carried out.

## Identification and Passports

In the UK, all births must be registered within twenty-seven days. The calf is issued with a passport, which records its unique eartag number, date of birth, sex, breed, eartag number of dam, and location of birth. This passport must accompany the calf throughout its life and all movements must be recorded and registered with the British Cattle Movement Service (BCMS). When the animal dies, its passport must be returned to BCMS.

## Euthanasia

From 1996 to 2006, the value of male dairy calves in the UK was so low that many were euthanased soon after birth. From time to time it may also be necessary to euthanase sick calves or those born with congenital deformity. Legally, calf euthanasia must be performed to conform with the Welfare of Animals (Slaughter or Killing) Regulations (1995), which states that 'no person shall engage in the … slaughter or killing of any animal unless he has the knowledge and skill necessary to perform those tasks humanely and efficiently,' and without causing any unnecessary suffering or distress. In practical terms, this usually means the calf must be shot or the veterinary surgeon must administer lethal injection (generally not done because of cost). Carcasses must not be buried, but disposed of by incineration.

If euthanasia is not carried out immediately after birth, the calf should receive colostrum and the same level of care as any

> ### Legal Aspects of Castration, Disbudding and Removal of Supernumerary Teats
>
> - In the UK, use of local anaesthetic is mandatory for castration of any calf over two months old. If a rubber ring is used for castration, local anaesthetic must be used if the calf is more than seven days old. Local anaesthetic must be used for disbudding and dehorning of any cattle. Local anaesthetic must also be used for removal of supernumerary teats from heifers more than three months old.
> - Only calves less than two months old may be castrated by anyone other than a veterinary surgeon. A vet must remove supernumery teats of calves that have reached the age of three months or more.

other calf to protect it from disease. This is partly to preserve its welfare but also because on some farms outbreaks of infectious disease (such as scouring) have started in male calves that have no economic value (and have therefore been relatively neglected), and then spread to valuable dairy heifer calves.

# COMMON DISEASES IN YOUNG CALVES
## Mortality

It is inevitable that occasionally a calf is born dead or dies within the first two weeks, but target mortality level should be 5 per cent, and if more calves are dying the problem should be investigated (*see* Chapter 9).

## Prevention and Control

Even where there has been no recent change in management to precipitate an increase in calf mortality, the following steps should be taken:

1. Improve hygiene around calving: calving pens; calf accommodation; feeding implements.

2. Use a stomach tube to make sure all calves receive enough colostrum.

3. Increase level of supervision at calving.

4. Correct any inadequacy of cow nutrition – extra feed for thin cows and supply mineral/trace element lick.

5. Seek veterinary advice if cows are too fat at calving – this is a difficult problem and certainly should not be rectified by suddenly reducing feed available to animals at a late stage of pregnancy.

6. Consider vaccination of cows to promote antibody levels in colostrum.

## Scouring (Diarrhoea)

This generally occurs as a result of inadequate colostrum intake, digestive disturbance or infection, or a combination of all three. The scouring calf rapidly becomes dehydrated and acidotic (the pH of the tissues falls), with loss of salts and low blood glucose; it is dull, stops feeding and may die. Prolonged scouring causes weight loss and ill-thrift, with scalding of the skin around the hind legs and tail. The scouring calf is susceptible to other disease (such as pneumonia) and fails to grow. Effective early treatment and control of risk factors are therefore essential.

*Treatment*
Treatment focuses upon replacement of lost fluid and salts using oral rehydration therapy (ORT) – by stomach tube if the calf won't suck. The calf's need for energy continues, so a small milk feed should be given between each ORT drink (four hours apart). Antibiotic cover may be required, so contact your veterinary surgeon. Veterinary treatment with intravenous fluid to correct dehydration and acidosis may be needed if the calf collapses.

*Prevention and Control*
1. Make sure all calves receive at least 3.5 litres of colostrum within the first six hours of life, and continue to be fed colostrum for

### Common Causes of Early Calf Mortality

| Cause of death | Factors involved |
|---|---|
| Dystocia (difficult calving) *(See also* chapters 2 and 7) | • Large calves (owing to choice of bull) and/or small cows<br>• Inadequate supervision – stockpersons too busy<br>• Insufficient skill to deal with difficult calving, or intervention too early or too late. It does not matter if there is insufficient skill, provided assistance is sought if calving is not progressing normally<br>• Cows/heifers too fat at calving<br>• High incidence of milk fever (so cows fail to expel calf) |
| Calves small or weak at birth, born dead or die within 24 hours | • Inadequate feed for cows/heifers prior to calving<br>• Trace element deficiency (particularly iodine, vitamin E/selenium)<br>• Infectious disease such as Neospora, BVD |
| Calves fail to suckle – become weak and die | • Inadequate supervision – stockpersons too busy<br>• Group size too large so mis-mothering occurs<br>• Inexperienced heifer that will not allow calf to suckle<br>• Older cow with large or engorged teats – calf cannot suckle |
| Calves die within a few days of birth, with or without clinical signs before death | • Usually the result of inadequate colostrum intake<br>• Poor quality colostrum (often owing to inadequate feed intake of cows)<br>• Colostrum does not contain the specific antibodies required (especially calves born to heifers or bought-in animals)<br>• Poor hygiene allows overwhelming bacterial infection |

at least twenty-four hours. This provides specific antibody protection against micro-organisms that cause scour in young calves (e.g. *E. coli*, rotavirus and coronavirus.)

2. Feed milk at exactly the correct specifications: make sure that milk is thoroughly mixed and made up to the same concentration and temperature every feed. Give the right quantity each feed. It is difficult to make sure that every milk feed is the same when feeding whole waste milk. It is impossible to feed the same quantity each time where calves feed on an ad lib system.

3. Improve hygiene. All utensils used for feeding calves should be thoroughly washed between each feed. Bedding in calving boxes and calf-rearing pens should be clean and dry.

4. Isolate scouring calves from healthy animals to prevent spread of infection.

5. Send samples of dung or a dead calf for post-mortem examination for veterinary investigation of the cause of scour, so that a specific control plan can be formulated.

## Navel Ill and Joint Ill

The calf suffering with navel ill has a swollen and painful navel, and may be dull and lethargic with a high temperature. Navel ill is caused by bacterial infection of the navel soon after birth, but clinical signs are often not seen until the calf is ten to fourteen days old. Occasionally navel ill is not recognized until the calf is much older, when it is seen as an abscess at the site of the navel, often associated with an umbilical hernia caused by infection.

Once bacteria have invaded the navel, they may circulate in the blood and settle in one or more joints to cause joint ill, in the liver to cause liver abscesses, or in the brain and spinal cord to cause meningitis. Joint ill causes swelling and lameness in affected joints, and meningitis causes nervous signs such as fits, inability to stand, uncontrolled 'paddling' of the legs, and so on. Liver abscess may cause acute disease and death, or the affected calf may simply fail to thrive and gradually waste.

*Treatment and Control*

Treatment of all these diseases requires broad-spectrum antibiotic, with anti-inflammatory agents, but calves may die despite treatment, and infected joints may never regain normal function. It is therefore important to prevent disease by maintaining excellent hygiene around calving, ensuring that calves receive plenty of good-quality colostrum, and dipping navels with tincture of iodine soon after birth.

*This dirty milk feeder is likely to harbour disease organisms that cause diarrhoea.*

# CHAPTER 4

# Heifer Rearing

Rearing heifers as herd replacements is an important component of farm management. These animals will be the future of the enterprise: good heifer rearing provides a sound basis for profitable cattle keeping, while failure to rear them appropriately may lead to economic losses. These losses occur either during the rearing period or after the animals join the herd, through increased mortality, increased disease, forced culls and reduced productivity. It is estimated that in the United Kingdom 20 per cent of heifers reared for dairy production never reach the end of their first lactation. Given that the cost of rearing a heifer in the UK is about £900 at the time of writing (2007), it can be appreciated that this wastage is extremely costly both in financial and in animal welfare terms.

Successful heifer rearing requires attention to detail. A good programme sets defined

*Home-bred heifers being reared as replacements for a suckler herd.*

targets and ensures that these are achieved through regular monitoring of the animals throughout the rearing period.

## TARGETS

The ultimate objective is to raise, at the least cost, an animal that, once calved, is capable of maximizing her innate genetic potential for production. Central to the cost of production is the chosen age for calving. For both beef and dairy farming the accepted optimal target is generally twenty-three to twenty-four months, although it is later for slower growing traditional and rare-breed cattle. Below this age there will be an increased incidence of calving difficulties and reduced productivity post calving. Heifers that calve late may become overfat before calving and rearing costs are increased. For example, calving at thirty months as compared to twenty-four months in the UK is considered to raise the rearing cost by 50 per cent. However, if the farm cannot achieve target weight gains required to calve at twenty-four months, it is better to calve later and accept increased rearing costs than to calve small, unproductive animals.

### Target Number of Heifers Required

*1. The number of cows and heifers that must be bred per annum to produce all required replacements to maintain herd size is a function of:*

- The average herd cull rate, which determines the number of heifer replacements required.
- The average calving interval, which determines the number of cows that must be bred to produce that number of heifers in one year.
- The average mortality rates of heifers from birth to calving, which determines how many extra live births are required to produce that number of heifers.

The number required is calculated as:

*Small heifer with large cows: heifers that calve in too small are unable to compete with the main herd.*

Expected % adult herd cull rate/100 × herd size
multiplied by (calving interval (days)/365)
multiplied by (100/100-expected % heifer mortality)

If no sexed semen is used, this result must be multiplied by two since half the calves born will be male.

Thus in a 200-cow herd with a 420 days average calving interval, a 30 per cent adult cull rate and a heifer mortality rate to calving of 15 per cent, the number of cows/heifers that need to be bred per annum to produce the required future herd replacements is:

$$(30/100 \times 200) \times (420/365) \times (100/(100-15) \times 2$$
$$= 10 \times 1.15 \times 1.18 \times 2$$
$$= 162$$

N.B. In a 200-cow herd with a 370 days average calving interval, a 22 per cent adult cull rate and a heifer mortality rate to calving of 5 per cent, the number of cows/heifers that must be bred per annum to produce the required future herd replacements is only 94.

*2. The total number of calves and heifers in the rearing group is a function of:*

• The average herd cull rate, which determines the number of heifer replacements required.
• The average age to calving, which determines how long each animal remains in the rearing group.

Expected % adult herd cull rate/ 100 × Herd size
multiplied by average age to calving (months)/12

In the example 200-cow herd – with 30 per cent cull rate and average age to first calving of twenty-eight months – the number of heifer replacements being reared at one time is:

30/100 × 200 × 28/12
= 140

This will be a minimum estimate as it has not taken into account those animals being reared that die before they calve. The actual number will be higher, depending on the stage of rearing that losses occur prior to calving.

## Targets for Growth

Heifer rearing is expensive, with 60 per cent of the costs being incurred in the first nine months. There is a temptation to find feeding and management systems that reduce rearing costs. However, it is no good saving feed costs for a calf if it later results in increased mortality, increased time to service, and smaller heifers at calving. Ensuring that target growth rate is met during this period is essential to maximize the future fertility and productivity of the animal.

The basic precept for both beef and dairy sectors is to ensure rapid and consistent growth rates, as the energy derived from feed is converted to growth more efficiently this way.

The target growth rate for heifers varies according to age and is influenced by the planned age and weight at calving and by breed. Target growth rates are subject to much research to ensure that they result in the most productive animal possible. The aim is to achieve a well-grown heifer at calving; in general the heavier the heifer at calving, the less growing she needs to do in first and second lactation, so the

### Table 6:   Average required daily weight gains for dairy breeds assuming a target of 24 months to calving

| Breed | Weight | Age (months): 0–3 | 3–6 | 6–12 | 12–16 | 16–24 |
|---|---|---|---|---|---|---|
| Friesian/Holstein | Liveweight gain (kg/day) | 0.85 | 0.85 | 0.8 | 0.85–0.9 | 0.7–0.75* |
| | Start weight (kg) | 40 | 100–120 | 180–200 | 320–340 | 420–440 |
| | End weight (kg) | 100–120 | 180–200 | 320–340 | 420–440 | 600–620 |
| Jersey | Liveweight gain (kg/day) | 0.65 | 0.65 | 0.65 | 0.7 | 0.65 |
| | Start weight(kg) | 20–25 | 70–80 | 120–140 | 220–240 | 300–320 |
| | End weight (kg) | 70–80 | 120–140 | 220–240 | 300–320 | 420–440 |
| Ayrshire | Liveweight gain (kg/day) | 0.85 | 0.7 | 0.75 | 0.75 | 0.7 |
| | Start weight(kg) | 30–35 | 100–120 | 160–180 | 290–310 | 370–390 |
| | End weight (kg) | 100–120 | 160–180 | 290–310 | 370–390 | 540–560 |

* Growth rates will reduce to 0.7kg/day in the last two to three months prior to calving

**Table 7: Average nutrient requirements for Holstein heifers to achieve target liveweight gain**

| *Weight* | Age (months): | | | | | |
| | *3* | *6* | *9* | *12* | *16* | *24* |
|---|---|---|---|---|---|---|
| Liveweight gain (kg/day) | 0.85 | 0.85 | 0.75–0.8 | 0.8–0.85 | 0.85–0.9 | 0.7 |
| Dry matter intake (kg) | 3.8 | 5.5 | 8 | 9 | 11 | 15 |
| ME (metabolizable energy) requirement per kg fed | 11.8 | 11 | 9.5 | 9.5 | 9.5 | 10.5 |
| CP (crude protein) requirement % (approx) | 16–20 | 16–18 | 16 | 14 | 12–14 | 16 |

**Table 8: Average required daily weight gains for beef heifers (autumn-born calves) assuming a target of 24 months to calving**

| *Breed* | *Weight* | Age (months): | | | | |
| | | *0–3* | *3–6* | *6–12* | *12–16* | *16–24* |
|---|---|---|---|---|---|---|
| British × Friesian | Liveweight gain (kg/day) | 0.6 | 0.7 | 0.6 | 0.6 | 0.6 |
| | Start weight (kg) | 30 | 90–100 | 150–170 | 260–290 | 340–370 |
| | End weight (kg) | 90–100 | 150–170 | 260–290 | 340–370 | 490–520 |
| Continental × Friesian | Liveweight gain (kg/day) | 0.85 | 0.7 | 0.75 | 0.75 | 0.7 |
| | Start weight(kg) | 30–35 | 100–120 | 160–180 | 290–310 | 370–390 |
| | End weight (kg) | 100–120 | 160–180 | 290–310 | 370–390 | 540–560 |
| Charolais | Liveweight gain (kg/day) | 0.85 | 0.85 | 0.8 | 0.85–0.9 | 0.75 |
| | Start weight(kg) | 40 | 100–120 | 180–200 | 320–340 | 420–440 |
| | End weight (kg) | 100–120 | 180–200 | 320–340 | 420–440 | 600–620 |

more milk she will produce either for rearing her calf or in the parlour.

However, heifers should not be grown as fast as possible all the time. It has been a long-held belief that fast growth rates in the pre-pubertal period result in excess fat deposition in the udder of dairy heifers, leading to reduced mammary tissue development and subsequent reduced milk production. It is now understood that it is not the rate of weight gain but whether the animal becomes too fat that determines fat deposition in the udder. Thus growth rates of up to 0.85kg per day for a Holstein can be maintained provided that the growth is directed towards lean tissue and not fat. This can be most influenced by providing suitable levels of protein and avoiding diets with a high starch and low protein content. Heifers should calve at condition score 3.0 and certainly no more than 3.5 (based on a condition scoring system of one to five, *see* Chapter 5). Target growth rates are laid out in Tables 6–8.

## NUTRITIONAL MANAGEMENT

### Criteria for Weaning

If weaning is performed incorrectly there will be a growth check that undermines the high growth rates in the pre-weaning period and, at worst, the stress can induce disease and mortality. There are different practices and ages for weaning heifers bred within the dairy sector, but the most common practice is to wean at six to eight weeks (*see* Chapter 3).

*Weaned calves. Suckler heifers are well grown at weaning: they must calve at about twenty-four months or they will become too fat.*

Heifer replacements bred from suckler cows are weaned at around seven to nine months of age. Weaning is now usually performed by preventing calves access to their dams but still allowing them sight and sound of them, as this is considered to reduce the stress associated with weaning.

If cattle are at grass a leader–follower grazing strategy can be used: calves are allowed access to the paddock prior to the cows, allowing them to select the most nutritious grazing.

## After Weaning

The rumen is not fully developed until four to six months of age. A correct balance of concentrate and forage is needed to ensure healthy rumen development. Excessive concentrate feeding, particularly in conjunction with poor intakes of forage, leads to bloated rumens and stunted rumen papillae (lining the rumen) owing to acidosis. Inadequate supply of concentrate leads to dull-coated, 'pot-bellied' calves.

At around ten to twelve weeks of age, concentrate is changed from an expensive coarse mix or calf pellet with 18 per cent crude protein (CP) to a 16 per cent CP pellet. From three to six months of age, housed heifers should be fed approximately 2kg of a 16 per cent CP concentrate containing a high proportion of undegradable protein, plus good-quality forage ad lib.

Weaned heifers from the dairy herd can be grazed from the age of three months, though supplementation with alternative forage, particularly in adverse weather conditions, is advisable to ensure intakes are maintained.

Many dairy farms rear heifers with a total mixed ration (*see* Chapter 5), similar to that fed to the low-yielding cows. This has the advantage of being less labour intensive. It also ensures that the concentrate component of the diet is taken in more slowly by all members of the group than it is when feeding concentrates alone.

From six months of age the proportion of concentrate and crude protein content of the diet varies according to the forage supplied. Forage quality should be analysed so that

*After weaning, heifers are reared in small groups of similar age. This reduces spread of infections such as pneumonia, and allows feeding to be targeted precisely to meet the group's needs.*

supplements can be accurately calculated. For example, good-quality grazing supplies almost all the nutritional requirements for heifers over nine months of age (with the possible exception of trace element requirements). Most forages (hay, silage, barley straw) are suitable and should be fed ad lib. If silage is fed it should be supplied fresh every day to avoid secondary fermentation and spoilage. Forage

intake is likely to be about 2kg of dry matter per 100kg bodyweight.

Following service, heifers can be fed as above until the last three months of pregnancy. At this point, owing to increased feed requirements and a relative fall in feed intake, the energy density of the feed should be increased. If inadequate energy is fed at this stage heifers are likely to be too small at calving,

*If heifers are yoked when they are fed they can be given routine treatments or pregnancy tested as required.*

*A weigh crush provides the easiest method of monitoring heifer growth. It should be used in conjunction with condition scoring.*

*As part of the process of monitoring heifer growth, a measuring stick can be used to determine height at the withers.*

while excessive supplementation causes excess fat. Both increase the risk of calving problems and subsequent poor production.

## MONITORING GROWTH AND CONDITION

There are two methods to monitor the growth of heifers:

1. The measurement of bodyweight using scales or a weighband.

2. The measurement of the height of the animal at the withers, which is a measure of the skeletal development of the animal.

Bodyweight should be used in conjunction with body condition score (*see* Chapter 5) to avoid the potential for assuming that a small fat heifer is meeting her target growth. For practical reasons scales are easier to use than other methods, though a practised eye can develop a good sense of likely wither height. Whatever the system adopted, growth rates should be monitored regularly, preferably every month, and plotted to check that the heifer is fulfilling the target requirements. This allows early detection and correction of problems.

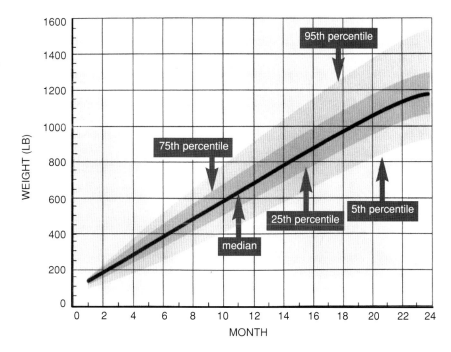

*Graph showing targets for growth in heifers (US Holsteins). (From Heinrichs, Pennsylvania State University.)*

It should be remembered that disease factors, for example a parasite burden, may also cause disappointing growth rates. After a period of poor growth heifers exhibit compensatory growth (*see* Chapter 5) when offered an energy-dense ration, which can allow them to catch up if deficits are caught early.

## HEIFER REPRODUCTION

Fertility management is as important as growth if heifers are to calve at twenty-four months. Management factors that must be considered are age, size and weight at first service.

### First Service

The heifer must be sufficiently well grown to calve normally and maintain herself in her first lactation. Generally the target is to breed when the heifer reaches 60–70 per cent of her estimated mature bodyweight. She should be in good condition – condition score 3 to 3.5 – and, assuming she is to calve at twenty-four months, she should be fourteen to fifteen months of age. (The exceptions to calving at twenty-four months are autumn-born calves destined for spring-calving suckler herds, which tend to calve at thirty months, and slow-growing native and rare

## Table 9: Targets for weight at service for a variety of breeds

|  | *Breed* | *Weight (kg)* |
|---|---|---|
| Dairy | Jersey | 250–280 |
|  | Ayrshire | 320–350 |
|  | Holstein | 370–400 |
| Beef | Hereford × Friesian | 330–350 |
|  | Continental × Friesian | 360–380 |
|  | 100% Continental Cross | 410–430 |

breeds, particularly those kept in a harsh environment.) The outline targets for weight at service for a variety of breeds are shown in Table 9 on page 73.

## AI or Bull

The decision to use a bull or artificial insemination (AI) – or a combination of the two – in the service of heifers will depend on the management systems and the breeding policy the farmer wishes to employ. The relative advantages and disadvantages of each technique are outlined in the box below.

## Selection of the Sire

The bull should always be chosen to produce offsring that are easy to calve. Test AI bulls, though cheaper, should not be used on heifers or there may be a risk of calving difficulties. (Test AI bulls are young animals for whom there is not yet enough data to determine the growth rates or milk yield of their progeny, nor whether their calves will be delivered easily.)

## Synchronization

Where AI is to be used, either alone or in conjunction with natural service, the artificial synchronization of the oestrous cycle using hormones can be useful. It allows heifers to be served at a fixed time without the requirement to look for the signs of oestrus. It also helps to ensure that heifers calve in a tight block at the desired time, thereby aiding their management. The optimal time to calve heifers is always at the start of any calving period so they have the maximal time to recover from calving in preparation for breeding again.

A variety of synchronization regimens can be adopted; all incur costs for drugs and veterinary input. These costs are outweighed by the benefits of reduced labour input and tight calving pattern, provided that pregnancy rates are good (approaching a target of 60 per cent). Heifers must therefore be well managed to optimize their potential fertility before entering a synchronization programme.

## Post Service

Dairy heifers in the latter stages of pregnancy (six to eight months) should be introduced to the post-calving environment – for example by running with the late-lactation cows – so that they can become accustomed to housing and the parlour. This reduces stress

---

### AI versus Bull

|  | Advantage | Disadvantage |
|---|---|---|
| *AI* | • Allows greater control over genetic selection, including ease of calving.<br>• Will allow faster genetic improvement of the herd.<br>• Avoids problems of inbreeding.<br>• Easier to control, manipulate and monitor fertility. | • Requires more labour and management input and infrastructure.<br>• Higher cost*. |
| *Bull* | • Low labour input.<br>• Optimum heat detection efficiency and multiple serves per heat should produce better pregnancy rates than the use of AI. | • Genetic potential often not known.<br>• More risk of disease.<br>• Bull has risk of problems with fertility.<br>• Risk of trauma to heifers.<br>• Health and safety aspects of working with a bull. |

* In smaller herds a bull may be more expensive to run.

*A bull chosen to run with heifers must be known to produce calves that are delivered easily, and he should not be too big or heavy.*

post calving. In the last two to three weeks pre-calving the heifers should be moved to the paddock or yard where they will calve. This not only allows the introduction of a suitable transition diet but also reduces stress caused by sudden changes in their environment close to calving.

## POST CALVING

Following calving, first-lactation dairy heifers should ideally be managed as a separate group. Trials consistently show that reducing the stress of competition with older animals results in increased production, reduced disease,

*Heifers that are not used to cubicles prior to calving may avoid cubicles and lie in the yards. They then get very dirty and are at risk of developing mastitis.*

*Well-grown heifers that have suffered no disease setback will reach target size/weight for service at fourteen to fifteen months, ready to calve at twenty-four months.*

and better heifer survival. Where this cannot be done it is important that heifers are placed in an environment with minimum competition for water, bedding and feed.

## CONTRACTS

Many enterprises opt to contract out the rearing of their heifers. This allows for increased management focus on the adult herd, and, by freeing up resources, can allow an increase in herd size of up to 25 per cent. Where heifer rearing is contracted out, a contract should be drawn up, outlining the targets, requirements (e.g. for vaccination or breeding), and responsibilities of both parties (including responsibility for costs and any losses that may occur).

## HEALTH AND DISEASE

The primary diseases that impact on heifers are respiratory and parasitic.

Pneumonia tends to be a group problem, and is usually first recognized as coughing. Affected animals have a fever, breathing may be laboured (dyspnoea), and there may be discharge from the nose and eyes. In dairy herds it usually affects housed calves aged four weeks or more, although occasionally it is seen in younger animals. In beef herds it is classi-

cally associated with housing. Severe disease causes widespread lung damage and, even if the calf recovers, lung function may never be fully restored: the calf remains stunted and poor, and likely to succumb to further bouts of pneumonia. It is a multi-factorial disease that occurs as a result of the interaction of infectious organisms, the environment and the innate resistance of the calf. These factors are discussed further in Chapter 9.

Parasites, specifically gutworm and lungworm (*see* pages 190–1), are a risk. Liver fluke is also a risk in some areas. The prevention of parasitic infestation requires:

- Grazing that harbours a relatively low parasitic burden (clean grazing) by using crop rotations or by grazing heifers on pasture after silage/hay has been cut.
- Grazing the land with adult stock either in conjunction with the heifers or prior to the heifers. Although the adults may eat infective larvae they do not usually show signs of disease because they have built up immunity.
- Strategic use of drugs to control parasites.
- Vaccination to prevent lungworm.

A control plan must be designed for the individual unit.

# CHAPTER 5

# Nutrition of Adult Cattle

Today's cattle have evolved from animals that fed on mixed, poor-quality grazing. The structure of plant cells, including grass, cannot be broken down by a simple digestive system – such as that of the human or dog – so cattle have developed the rumen: a huge, unwieldy and biologically complex organ that allows ruminants to extract energy and protein from otherwise undigestible plant material. Extensively managed beef cattle continue to thrive on a diet based solely or mainly on grazing, but the production demands imposed on the intensive beef fattener and the high-yielding dairy cow cannot be met by poor-quality grazing. They must be fed a very different ration which, potentially, does not suit normal rumen function. A balance must therefore be struck between meeting the animal's high nutritional demands and maintaining or promoting a healthy rumen.

This chapter explores how that balance can best be achieved. The discussion is applied mainly to the dairy cow because she is the animal (with the exception of cereal beef) most at risk of nutritional imbalance, but the principles are the same for all.

## PRINCIPLES

### The Digestive Tract

The tongue is thick, rough and muscular for tearing off grass. There are eight lower incisors but no upper incisors; instead there is a dental pad of thickened tissue. The molars provide a flat grinding surface for breaking down forage.

Food passes from the mouth to the stomachs via the oesophagus. The bovine stomach system is divided into four parts, of which the first, the rumen (diagrammatically illustrated on page 78) functions as a fermentation vat where microflora, specifically protozoa and bacteria, digest plant material. The rumen has a muscular wall that contracts to churn over its contents regularly, once or twice per minute. Gas produced as a by-product of rumen fermentation is belched up as the rumen contracts. The ruminant regurgitates semi-digested feed from the rumen to be chewed a second time (known as ruminating or chewing the cud). This allows plant-cell structure to be further broken down and mixed with saliva before being returned to the rumen for more fermentation. Some products of fermentation are absorbed directly from the rumen, but microbial cells and the remaining undigested energy, protein and fat pass through the next two stomachs, the reticulum and omasum, and on to the abomasum or 'true stomach'.

From this point, the digestive tract is similar to that of a single-stomached animal. Plant material broken down in the rumen is digested into its constituent parts and absorbed into the bloodstream from the small intestine. The main function of the large intestine is to absorb water and minerals from the gut content so that semi-solid dung is excreted from the rectum.

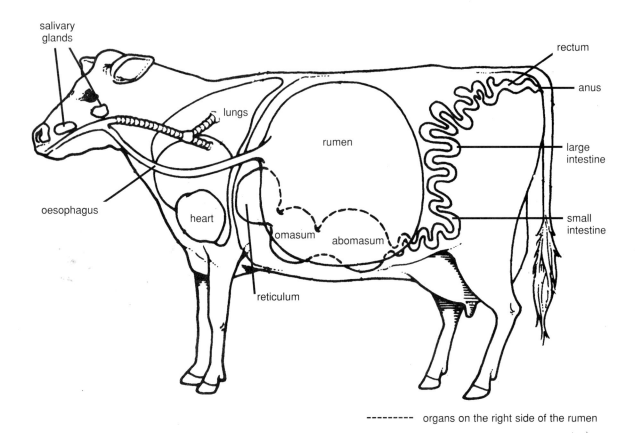

-------- organs on the right side of the rumen

*Schematic view of the adult ruminant abdomen.*

## The Function of the Rumen

The aim of feeding is to supply the animal with a ration that supplies its nutritional requirements, but this must be done in a way that promotes a good fermentation response in the rumen.

The main end-products of fermentation are the volatile fatty acids (VFAs), especially acetic, propionic and butyric acid, and microbial cells. The VFAs act as the major energy source for the cow and are absorbed through the rumen wall. In the rumen, nitrogen sources, both protein and non-protein, are broken down to yield energy in the form of VFAs and basic nitrogen compounds, particularly ammonia. These nitrogen compounds are used as building blocks for microbial protein that passes into the gut to be digested to release amino acids. Not all the ammonia is utilized, and the excess passes through the rumen wall and into the bloodstream, where it is metabolized by the liver into urea. The conversion of ammonia into urea costs energy (termed the urea cost), so excessive ammonia passing through the rumen wall is undesirable.

The proportion of different end-products produced by the fermentation is influenced not only by the physical and chemical properties of the feed, but by the fermentation conditions in the rumen. Maximizing the efficiency of the rumen is based upon two precepts common to all fermentation processes:

1. Ensuring that the microflora receive the correct raw ingredients to thrive and survive; and

2. Maintaining a stable environment, particularly the level of acidity.

In practical terms this means that the supply of energy and protein must be sufficient to meet microbial requirements and that the fermentation of the feed – particularly the carbohydrates – must occur in a controlled manner.

Controlled fermentation is important because rapid fermentation of sugars and starch is accompanied by the production of acid. The more acid that is produced the lower the rumen pH. As pH falls, the population of microflora in the rumen is altered, favouring organisms suited to a more acid environment. The net result is a reduction in the efficiency of the rumen, particularly in the ability to digest fibre, so that the cow's appetite and level of production falls. Further reduction in pH creates a more hostile environment, and in severe cases, where excessive levels of highly fermentable feeds are eaten, the rumen pH may fall to the point where the cow becomes ill and may die.

The cow's main defence against the development of an acid environment in the rumen is to neutralize the acidity with bicarbonate, which is produced in saliva. This is termed rumen buffering. Saliva is secreted and mixed with the feed when the animal chews, either during the initial feed intake or when feed is regurgitated and rechewed during rumination. This is vital to the maintenance of a stable, healthy rumen. It follows that feeds requiring a lot of chewing – for example those that contain fibre, are dry and have a large particle size – promote rumen buffering. This is generally the case with forages and hence it is important to ensure there is sufficient forage in the diet. The fibre content of forages varies tremendously – for example, straw contains a much higher level of fibre than grass. The particle size of conserved forages also varies, depending on how they were processed during harvesting. Forage that has been cut into very small particles (short chop length) provides less rumen buffering than long-chopped forage.

## FEEDS

There is a massive variety of feeds for cattle. There are also many feed additives available to enhance rumen efficiency and cow productivity. However, there are two major categories of feed: forage and concentrates. Forages are feeds composed primarily of plant cell-wall material while concentrates are primarily composed of protein, sugars and starch. Forages are cheap and important for rumen pH stability; concentrates are useful to increase the energy and protein content of the diet and to increase intakes, but they are generally expensive and may promote rumen acidosis.

Within these groups the feeds can be further subdivided on the basis of their relative energy and protein content. Further divisions can then be made on the relative rumen degradability and speed of degradability of these nutritional components.

The nutritional value of feeds depends upon:

- Dry Matter (DM) content. Feeds contain a proportion of water that may vary from less than 10 per cent in rolled barley or straw to 90 per cent-plus in lush spring grass.
- Fibre content. As discussed above, fibre content is an important determinant of rumen function.
- Energy, protein (and fat) content. Feeds broken down by microflora in the rumen vary in the nature of the end-products of fermentation and the speed at which they are broken down. For instance, feeds high in starch and sugars (for example, grains such as wheat) will be far more rapidly broken down into basic carbohydrates than feeds high in cellulose (such as straw). Both are inherently low in nitrogen, so yield little nitrogen for protein production. Although most feed is broken down by microflora in the rumen (the rumen degradable energy and protein), some passes through the rumen to be digested further down the digestive tract (undegradable energy and protein). Fats have the advantage of being

---

### Feed Analysis

Laboratory techniques are used to analyse feeds. Analysis determines:

- Dry Matter content.
  The water content of the feed does not contribute to its nutritional value, so feed analysis is based upon its DM content (total mass of feed less water content).
- Fibre.
  This can be measured in a variety of ways. The commonest is to measure the proportion of feed containing neutral detergent fibre (NDF), a measure of plant cell-wall content.
- Energy and protein.
  Analysis of feeds provides information on the proportions of degradable and undegradable energy and protein and, where degradable, the proportions that are degraded quickly and slowly.
- Micronutrients.
  'Book values' can be used for feeds whose micronutrient content is constant, but those whose vitamin and mineral content are variable must be analysed to find accurate values.

---

very energy dense, but unfortunately they will disrupt rumen digestion if fed in excessive quantities. They therefore need to be fed in a manner that protects them from rumen breakdown (so-called protected fats) if they are to be used to correct potential energy shortfalls.

- Vitamin and mineral content. The micronutrient content of some forages, such as maize silage and wholecrop, and most individual feeds (termed straights), are relatively constant. Other feeds, in particular grass-based feeds, will vary enormously in their vitamin and mineral content.

## Forage

Maximizing the quality and quantity of home-grown forage is an important component of farm profitability. Forages can be fed either in their fresh state or in a conserved form (usually as silage, hay or haylage). The use of grazed grass, by far the cheapest form of for-

age, is discussed in Chapter 6. Silage is produced by cutting the fresh crop and storing it either in a covered pit (the silage clamp) or in plastic-covered bales. It is anaerobically fermented or 'pickled'. This results in the production of acid which, when it has been produced in sufficient levels, stops the fermentation process, thereby preserving the crop. Failure of the fermentation process to produce sufficient acid, or ingress of oxygen to the crop, results in continued fermentation and therefore continued degradation of the nutrients within the crop. This will result in a poor feed value and a reduced palatability (owing to breakdown of protein into ammonia-based compounds). The requirements for the production of silages are important but are beyond the scope of this book. However, quality silage, whether it is grass, maize or wholecrop, is characterized by:

- Preservation of the highest proportion possible of the original nutritional properties of the fresh crop.
- Stability (whether the forage quality remains stable or is subject to secondary fermentation or spoilage from the growth of moulds).
- Palatability.

## Concentrates

Concentrates can be fed alone as a 'straight' (*see* above), or as a 'mix' (a combination of usually unmilled straights), or as 'cake/compound', which is a manufactured, and usually pelleted, combination of milled straights with a set protein and energy level.

Table 17 at the end of the chapter demonstrates the feed properties of some commonly encountered forages and concentrates.

## FEED REQUIREMENTS

Feed is required for:

- Maintenance of normal body function and health to maintain the status quo (maintenance requirement 'M').

*Beef cattle can often meet their nutritional needs on a forage-based diet because their milk yields are much lower than those of dairy cows.*

*Red Devon cows eating haylage at a round feeder. These animals are in excellent condition without supplementary feed.*

*Feed stored in silos like this is protected from vermin and weather, so its quality is maintained.*

- Lactation. The energy, protein and other nutrients required to produce every litre of milk. A diet set for M+30, meets the physiological requirements of a cow producing 30 litres of milk per day.
- Growth. Heifers that calve in at two years old continue to grow throughout first lactation.

- Pregnancy. Requirements for pregnancy are low until the last four to six weeks.
- Change in body condition.

For each of these processes the cow has requirements for macronutrients (carbohydrates, protein and fats) and micronutrients (minerals and vitamins). If we know her physiological requirements, how much she can eat, the nutritional properties of the feed and their effect on rumen function, we can formulate a diet to meet her requirements.

## FEED INTAKE

The amount an animal will eat is a complex issue. It is mainly determined by the animal's bodyweight and its requirements. For the lactating cow, feed intake can be estimated by the equation (0.025 × liveweight [kg]) + (0.1 × milk yield [kg]). Feed intake is also influenced by rumen fill, which in turn is determined by how long feed remains in the rumen. An animal will eat less of a ration that contains a high proportion of fibre (such as straw) – which is less degradable and stays longer in the rumen – than a ration that is rapidly broken down, such as spring grass. Feed intake falls around calving and takes three to four weeks to recover. In the dairy cow this coincides with a rapid increase in milk production, so feed intake will not meet her nutritional requirements. She draws on body reserves to meet this shortfall and loses weight (termed negative energy balance or 'milking off her back').

Feed intake should be maximized for maintenance of health and healthy rumen function. This minimizes the likelihood of energy deficit and also allows the feeding of a ration that has as low an energy density as possible for a particular production demand. Rations with lower energy density promote healthier rumen function and are cheaper. Excessive condition loss in early lactation will affect the cow's health and productivity, and should be prevented by, preferably, maximizing feed intake rather than by increasing energy den-

*Feed intake falls at calving. The rumen of this newly calved cow is empty. Note the hollow behind the ribs and below the spine, where the rumen lies.*

sity. Disease and any disruption of rumen function should be prevented, or treated early, so that feed intake is maintained.

## Rationing Adult Cattle

The development and feeding of a suitable ration for a herd can be divided into a number of key sections:

- Choosing the target level of production.
- Choosing the appropriate nutritional system.
- Assessing the type, quality and availability of farm forages.
- Balancing the ration.
- Feeding the ration.
- Monitoring the ration.

*Choosing the Target Level of Production*
This is discussed in full in the chapters on dairy and beef systems.

*The high-yielding dairy cow must consume as much as possible to prevent energy deficit.*

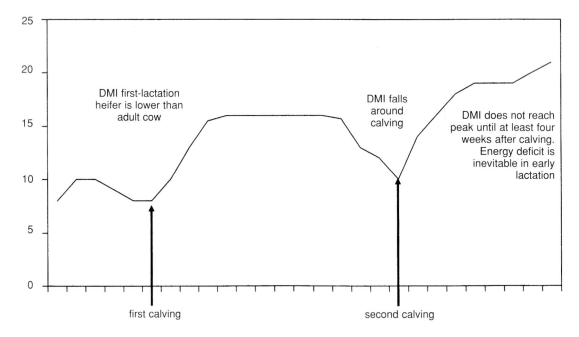

*Graph showing dry matter intake (DMI) of dairy cows at different stages of production.*

## Choosing the Appropriate Nutritional System

The next step is to choose the best feeding system. This section is predominantly aimed at the dairy farmer, since beef systems tend to use single forages, supplemented (when appropriate) with concentrate. However, where mixed rations are used on beef units, the same factors apply as for dairy cattle.

### TOTAL MIXED RATION (TMR)

Total mixed rations allow the feeding of both forage and concentrate together. This has the great advantage of stabilizing the rumen environment, thereby increasing intakes, by avoiding the introduction of boluses of highly fermentable concentrates (as occurs with parlour feeding), which tend to increase the risk of rumen acidosis. The system is expensive to operate and consequently most suitable for herds producing high yields. There are several different approaches to the TMR system (*see* Table 10).

### IN-PARLOUR FEEDING

In this system, some or all of the concentrates are fed in the parlour and the forage fed outside. The advantage is that feed costs are kept low by matching concentrate use to the individual cows. However, there are several disadvantages:

- There can be a tendency to overfeed cows in the parlour – total TMR removes this.
- It is very difficult to feed the high-yielding cow without risking acidosis and/or underfeeding. No more than 4kg of cake at one feed should be offered.
- It incurs compounding costs.

### OUT-OF-PARLOUR FEEDERS

This system can be used as part of a semi-TMR system or in combination with forage only. Feeders must be checked regularly to ensure they are operating correctly and that all cows are eating their quotient. Out-of-parlour feeders give cows more frequent access to

## Table 10: TMR systems

| TMR system | Advantages | Disadvantages | Notes |
|---|---|---|---|
| **One-group TMR**<br><br>The whole herd is fed as one with the same TMR. It works on the principle that the cow's intake broadly mirrors her requirements, so the high-yielding cow eats more than the low-yielding cow. | • Excellent level of control of the feeding, , as the effect of ration changes are reflected across the whole herd.<br>• Maximizes the opportunity to source cheaper feeds (straights) and makes best use of the capital outlay and extra labour involved in mixer-wagon operation<br>• There is only one group to manage.<br>• Cows maintain higher yields for longer especially where exclusively Holstein genetics are used | • Purchased feed rates (i.e. ration fed per unit milk produced) can rise if yield falls<br>• Eliminates the option to target-feed high-cost products (yeasts, buffers, protected fats) to higher yielding cows<br>• Risk of overfeeding cows in late lactation if peak yields are not high, if calving interval is prolonged, or if there is a spread of genetic potential within the herd<br>• Risk of underfeeding high yielders, leading to negative energy balance. Cows must be monitored closely<br>• Expensive | Maximum profits if:<br>• A tight calving interval (less than 410 days). Freshly calved cows (less than 150 days) are the most efficient in terms of feed conversion, so if the herd averages over 200 days in milk all year round the system will not be economic<br>• Good forages and/or very cheap concentrates (straights and/or by-products) |
| **Group TMR**<br><br>Different mixes are produced and fed to groups of cows based on their physiological needs. High-yielding cows receive a more energy-dense diet than low yielders. Group size is determined by housing, feeding facilities and parlour capacity. It should be possible to milk each group in no more than 1.5 hours | • Allows the feed rate to be controlled. High value components can be targeted to the appropriate cows. | • Too great a difference in nutritional density between the rations for different groups (there can be up to 50 per cent differences in the concentrate DM in UK diets). This can lead to poor persistency of yields, particularly after transfer to the low yielding group.<br>• Expensive to operate<br>• Complex to manage | • Best suited to herds that calve throughout the year<br>• Nutritional difference in concentrate DM between groups should be less than 10 per cent to avoid excessive yield drops.<br>• Higher-yielding herds should be divided into three or more groups because with just two groups the transition from one diet to another can be too great |
| **Semi TMR**<br><br>This is the most common feeding system in higher-yielding herds in the UK. It is strictly not a true TMR as it combines a mix with some level of in or out of parlour feeding of concentrate | • Has some of the benefits of TMR feeding while keeping the feed rate even tighter; it is therefore useful when trying to increase yields but control costs, or as a transition to total TMR feeding | • Increased risk of acidosis, especially in freshly calved cows whose DM intakes are still depressed. Every kg of cake fed in the parlour substitutes for a kg of more rumen-friendly mix; therefore acidosis is very likely.<br>• It requires discipline to avoid overfeeding concentrate in parlour | Parlour feed should increase slowly, by 0.5kg a day, to two-thirds full rate by fourteen days and full rate by three weeks post calving |

*A mixer wagon is required to feed TMR. The various components of the ration are measured or weighed and mechanically mixed before feeding.*

*In-parlour feeding enables rations to be matched to individual feed requirements. These computerized feeders automatically deliver the correct quantity to each cow when her number is entered.*

## Table 11: Comparison of forage feed values and cost

| Forage | Maize | Alkalage | Wholecrop s.wheat | Lucerne 4 years | Grass silage | Fodder beet | Dried wheat |
|---|---|---|---|---|---|---|---|
| Energy+protein cost (pence/litre) | 7.33 | 5.97 | 7.67 | 3.67 | 7.26 | 10.45 | 6.88 |
| Net cost of growing (£/acre) | 254 | 268 | 233 | 230 | 293 | 347 | 204 |
| Dry matter (tonnes/acre) | 4.8 | 4.9 | 4.0 | 6.0 | 4.5 | 5.8 | 3.0 |
| Fresh weight cost (£/tonne) | 15.88 | 35.73 | 23.30 | 11.50 | 16.28 | 10.84 | 58.29 |
| Dry matter cost (£/tonne) | 52.92 | 54.97 | 58.25 | 38.33 | 65.11 | 60.24 | 67.77 |
| Dry matter cost (pence/kg) | 5.29 | 5.5 | 5.83 | 3.83 | 6.51 | 6.02 | 6.78 |
| Energy cost (pence/megajoule of metabolizable energy) | 0.48 | 0.52 | 0.55 | 0.37 | 0.6 | 0.51 | 0.5 |
| Protein cost (pence/gram CP) | 0.06 | 0.04 | 0.06 | 0.02 | 0.05 | 0.1 | 0.05 |
| Energy cost (pence/litre) | 2.55 | 2.77 | 2.94 | 1.93 | 3.20 | 2.68 | 2.64 |
| Protein cost (pence/litre) | 4.78 | 3.19 | 4.73 | 1.73 | 4.07 | 7.77 | 4.24 |

*(Table courtesy of Kite Consultancy LLP.)*

cake than in-parlour only, so smaller amounts are fed at one time, which reduces the impact on rumen pH, the risk of acidosis and depression of dry matter intakes. It also allows a very controlled level of feeding. The disadvantages are that they are expensive to install and may be difficult to manage during the summer when cows are at grass.

*Assessing the Type, Quality and Availability of Farm Forages*

FORAGE INTAKES

The quantity that cows are eating must be estimated when cattle are grazing, but can usually be established fairly accurately when cows are housed. The intake of grazing cattle varies between 6.5kg and 20kg forage DM, depending on forage palatability and mix. Average intake is between 10.5 and 12.5kg DM.

FORAGE AVAILABILITY

Forage production depends upon crop yields, costs and forage characteristics. Examples of the approximate yields and costs for a variety of common forages are given in Table 11. They can be used to calculate the amount of forage the farm can produce. Quantity of forages already available can be calculated using the figures in Table 12, based upon clamp capacity and forage dry matter.

FORAGE QUALITY

Having established the quantity of forage that is required and available, the next step is to assess its characteristics. This can be done both by simple on-farm assessment and by chemical analysis at the laboratory. Standard laboratory analysis is cheap but is not totally reliable because there is variability in the analysis, and because forage may vary significantly at different points in the clamp. Analyses should be rechecked by submitting new samples regularly and by on-farm assessment.

*Fibre content and chop length* The forage must be inspected and handled to assess chop length and level of effective fibre. Both are implicated in the promotion of rumination and maintenance of a stable rumen pH. Forages with a very short chop length or that feel soft rather than spiky are unlikely to promote rumination. Opinions on the optimum chop length vary: a target of 19mm for maize is reasonable, though this may be lower if the crop is harvested in a relatively dry condition. The absolute minimum for wholecrop should be 19mm, although unfortunately this tends to be the maximum to which most contractors will go. Achieving a longer chop length than this for wholecrop often involves the removal of knives from the forage harvester. At the longer chop lengths recommended here, particular attention must to be paid to clamp consolidation at harvest to prevent aerobic spoilage and secondary fermentation.

*Dry matter* An accurate assessment of forage dry matter is important. This is because feed is delivered to the cow by measuring fresh weight. The fresh weight supplied is based on the assumed dry matter of the forage. If dry matter estimate is incorrect this can make a huge difference to the ration supplied, altering potential intakes and overall chemical and physical composition of the diet.

For example if the ration requires 6kg dry matter of a grass silage to be fed to 200 cows as part of a mix, and the dry matter of the silage is assumed to be 30 per cent, the farmer would expect to add $200 \times 100/30 \times 6 = 4000$kg of silage to the mixer wagon. If dry matter was in fact 25 per cent the real level of dry matter he has fed is $(4000/200)/(100/25) = 5$kg per cow, a deficit worth around 2 litres of milk. This commonly occurs where clamps are outside and the face becomes wet after rain.

*An estimate of silage dry matter content can be made by squeezing it.*

### Table 12: Weight of silages at different dry matters

| Dry matter content (%) | Weight (tonnes) per $m^3$ | | |
| --- | --- | --- | --- |
| | Grass silage | Maize silage | Grass silage (big bale) |
| 20 | 0.73 | | |
| 30 | 0.65 | 0.62 | 0.38 |
| 40 | 0.61 | 0.59 | 0.33 |

Dry matter of forage can be assessed by chemical analysis or on farm by:

- Squeezing the silage between the fingers and gauging the level of moisture released.
- Weighing the sample then heating in a microwave or on low oven setting until its weight becomes constant. Weigh again and divide by the original as a percentage.

The DM you see on silage analyses should always be checked by visual assessment or oven DM testing.

*TMR physical characteristics ('openness')* A healthy ration is generally 'open', i.e., it springs open when squeezed. This reflects its effective fibre content ('scratch') and dry matter percentage. A ration less than 40 per cent fibre tends to ball slightly.

*Spoilage and aerobic stability* There is no substitute for physically inspecting the silage clamps and feed troughs to ensure that the feed is stable and not fermenting. Where spoilage occurs it is characterized by production of heat and is often accompanied by a foul butyric smell and evidence of mould formation. It indicates that the quality of the feed is deteriorating, it reduces the palatability of the feed and predisposes to the production of toxins from moulds that may have a detrimental effect on the animals being fed.

Aerobic spoilage occurs where:

- There is air (by definition).
- There is a plentiful carbohydrate source.

*Silage clamps. Well-managed silage faces for maize and grass silage.*

- pH is greater than pH 4.5 (this tends to be in drier crops).
- There is a heavy inoculation of yeasts and moulds.
- There is poor attention to detail in managing the face of the silage clamp.

Aerobic spoilage may reflect an unstable crop, and there may be little that can be done to prevent it. However, the effects can be minimized as follows:

- Do not use visibly spoiled feed.
- Ensure that management of the face of the silage is kept as tidy as possible, for example by using a sharp shear grab to prevent air penetration deeper into the clamp.
- If the spoilage occurs predominately at the face, the feed rate should be increased by feeding to other stock or changing the ration formaulation, so that forage is fed before it deteriorates.

- If spoilage occurs in the trough, the trough should be cleaned daily. Feed should not be supplied in excessive quantities.

## Balancing the Ration

### Dairy

Ration formulation is usually performed by a trained nutritionist since understanding of cow nutrition has become increasingly complex. The principles of ration formulation are to:

- Maximize dry matter intake (DMI).
- Prevent acidosis.
- Maximize energy density without affecting the above.
- Maximize metabolizable protein yield while reducing the 'urea cost' to a minimum.
- Prevent metabolic disorders owing to mineral imbalances.
- To achieve the above at best cost to the farmer.

The process of formulation, based on a UK semi-TMR system, should follow the following protocol:

1. Choose the level of forage to be fed, based on the available quantities and the likely intake based on current and previous performance. For instance, each cow to receive 11.5kg DM.

2. Choose the appropriate maintenance plus yield (M+ figure) for the mixed part of the ration (this is usually based around the average daily milk production per cow). If cheap concentrate is available, or it is difficult to feed larger quantities of concentrate in the parlour, higher rates of feed may be fed through the trough. If a lot of cows are producing very low yields (perhaps owing to poor fertility, lameness or general management), it may be appropriate to feed less than maintenance plus average yield.

3. Make up the remaining shortfall by rationing cake appropriate to yield in the parlour.

To be able to formulate it is important to have an estimate of intake. This can be determined quite accurately where cows are fed in high- and low-yielding groups. Where cows at all stages of lactation are grouped together, the average intake of forage DM, plus any TMR concentrate DM, plus the average daily parlour feed, can be calculated using monthly tonnage divided by cow numbers and days in the month. Cows at peak lactation eat 10–15 per cent more than this average figure; low-yielding cows eat 10–15 per cent less.

For example, if average intake is 21.5kg DM (typical high-yielding level), the high-yielding group will eat:

21.5 + 10–15% kg DM
= 23.7–24.7 kg DM.

Some individuals will eat 28kg-plus and reach 60 litres, while others will eat only 19kg; but the ration is formulated to satisfy the average peak cow in the herd.

*Beef Fatteners*

There are two new concepts related to the growing animal:

- The efficiency of utilization of metabolizable energy for growth is influenced by the energy density of the ration.
- The efficiency of utilization of metabolizable energy for maintenance and growth is influenced by the rate of growth of the animal.

Rate of weight gain and energy density of the diet influence each other, so one or the other must be chosen to formulate the diet as follows:

- Determine daily liveweight gain.
- Establish the feeds to be used and their metabolizable energy and protein levels.
- Predict the energy density of the final ration.
- Determine the metabolizable energy and protein requirements needed in the ration.
- Predict the dry matter intake from the forage and the level of concentrate to be supplied.
- Formulate the ration.

Tables are available to perform this series of calculations manually but computer rationing systems perform these calculations without the heartache! A single diet formulation for an animal should not be for more than a total weight gain of over 100kg or to last for more than three months because it will no longer be accurate for the larger, heavier animal. After this point you will need to reformulate. Energy densities of the ration can be altered to allow adjustments to the target slaughter weight and age and to change the level of forage being used in the diet. For example, if forage stocks are low, provided there is no danger of fattening too quickly, an increase in concentrate use will not only reduce the level of forage consumed per day but also reduce the number of days it is fed because the animal reaches target slaughter weight sooner.

### Suckler Cow

The formulation of a ration for a suckler cow follows exactly the same process as for the dairy cow outlined earlier, though of course lactation demands are very much less and can only be estimated.

## Feeding

Maximizing dry matter intake should be a primary goal of every farmer. Not only does this allow higher yields to be produced but it also allows for reductions in the energy densities of rations for any given yield. Frequently too much attention is paid to formulating the ration on the computer and too little to the practicalities of cow and feed management.

### Cow Behaviour

Cows, like us, prefer to set their day out in a pattern: they like to have dedicated times for eating, drinking and ruminating. On average a dairy cow producing around 30 litres will want to have eleven half-hour meals a day, fourteen half-hour ruminating bouts, and make fourteen visits to the drinking trough. They prefer to do these activities in the sight

*Beef fattening cattle eating a trough-fed mixed ration.*

of other cows, whose eating acts as a stimulus for them to eat.

Cows are also hierarchical and, where there is competition for feed, water or bedding, sub-dominant cows will tend to lose out. When the make-up of a group of cows is changed, the cows will take time out from normal behaviour to sort out their new hierarchal order;

*Plenty of trough space must be provided to allow cattle to eat together, as they prefer. This prevents small cows from being bullied away from the trough. (Photo: Richard Vecqueray.)*

*A poorly designed feed trough with cows feeding head to head and one animal dominating a large area.*

most of this disruptive behaviour lasts for the first forty-eight hours.

All these factors are important in terms of the management of the cow. The impact of systems that fail to accommodate these behaviours, or cause excessive social disruption due to frequent group changing, are consistently underrated in terms of their effect on the cows, their feed intake and their consequent productivity. For example, when cows are denied access to feed they will try to make up the shortfall, not by eating for longer but by

eating faster; this is termed slug feeding. It is important that slug feeding is avoided because it increases the likelihood of digestive disorders and acidosis, particularly where there is a high concentrate-to-forage ratio in the diet or there is a lack of intake of long fibre.

Minimizing the impact of group changes is important, particularly in the period around calving (termed the transition period) where the cow or heifer is most susceptible to any fall in dry matter intake. The number of group changes should be kept to a minimum, but the following strategies can help reduce the impact of disruptive behaviour when making alterations to groups of cows is necessary:

- Avoid multiple small changes; make a smaller number of large changes and perform these changes on a routine basis.
- Keep groups of cows that will be mixed close to each other; for instance, have the freshly calved group next to the high-yielding cows.
- Make the move following afternoon/evening milking.
- As discussed in the heifer-rearing chapter, introduce heifers to the milking cow environment prior to calving.
- In the milking herd keep heifers as a separate group.

*Physical Facilities*

FEED TROUGHS

Given that cows are grazing animals, it should come as no surprise that they prefer to eat accordingly, with their heads down; so their feed should be set at 10–15cm above foot height. This also has the advantage of stopping a dominant cow controlling too much of the trough. (A rail or tensioned wire fence line placed 1.2m above foot height will further prevent this problem of individual cows dominating feed space.)

Similarly, cattle do not graze head-to-head, and when we force cows to eat this way – by giving them access to both sides of a narrow feed trough for instance – eating time will be significantly reduced (by up to 25 per cent)

| Table 13: Suggested dimensions for feed troughs | | | |
|---|---|---|---|
| Age (months) | Weight (kg) | Throat height (cm) | Neck rail height (cm) |
| 6–9 | 150–250 | 36 | 70 |
| 9–12 | 250–300 | 40 | 75 |
| 13–15 | 300–350 | 44 | 85 |
| 16–24 | 350–550 | 48 | 100 |
| Cows | 550–700 | 52 | 120 |

compared to a fence line feeding system. If cows are to face one another they should be at least 3m apart.

The trough surface should be smooth, light and clean. The smell of decomposing vegetation is not conducive to a cow's appetite and troughs with feed that is likely to be susceptible to degradation should be cleaned out daily. Trough space of 0.75–1m per cow is required to avoid reduction in intake for the more submissive cows.

WATER

High-yielding cows have a vast requirement for water for maintenance and production, and even a small limitation in water intake will decrease dry matter intake. Housed cows should have a minimum of two water sources per group. The troughs should be positioned close to the feed, out of the sun, and should not be more than 18m away from any cow. They should not be at the dead end of an alley; this can result in dominant cows protecting and limiting access to the trough. There should also be adequate open space around the water source. Alleys between cubicle rows should be at least 4.5m wide, to provide room to drink and sufficient space to allow cows to pass each other.

Water troughs should be located in the loafing area post milking. (Cows drink 50 to 60 per cent of their total water intake post milking.) Linear space should be sufficient to provide 30–60cm of trough space per cow. A good guideline is to provide 70cm of trough space per cow exiting the parlour. For a twenty-unit parlour with ten units per side, 7m of trough space should be provided.

Water pressure should be adequate to avoid cows having to wait for a drink. Troughs should be no deeper than 30cm to reduce water stagnation and to aid cleaning. Water

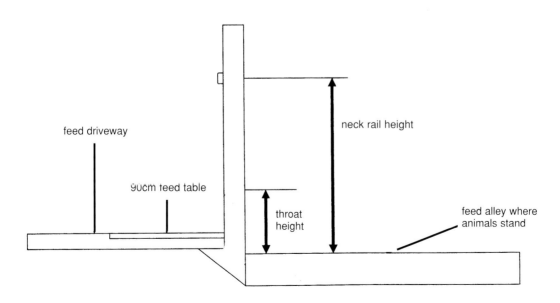

*Recommended feed-trough design.*

*This water trough is overcrowded: there is inadequate space for the animals to access water.*

quality can be evaluated for its smell, taste, physical and chemical properties. More simply, would you be willing to drink from the trough?

CUBICLE DESIGN

Cow comfort is an important aspect of feed intake and digestion. A cow consuming 25kg of dry matter should be spending about twelve and a half hours a day chewing her cud in a comfortable lying position. When more than one third of the cows in a herd are found standing or lying in alleyways, the design of their cubicles should be questioned. (This topic is discussed in greater detail in Chapter 8.)

PASSAGEWAYS AND LOAFING AREAS

Passageways should be wide enough to allow two cows to pass each other easily when walking behind the line of cows eating. All walking and loafing areas should be clean and non-slippery.

*Feed Preparation and Supply*

It is said that a cow receives four rations:

1. The one that is initially formulated.
2. The one that is subsequently prepared and offered to the cow.
3. The one that the cow then opts to eat.
4. The one that is actually digested.

Despite the scientific and technological advances that have enhanced our understanding of ration formulation, there is plenty of room for error if preparation and delivery of the feed is not performed correctly. The following should reduce the risk of errors:

*These cows have sorted through their ration, leaving the chopped straw which should stabilize rumen function.*

• Ensure that the person performing the feeding is well trained and understands the importance of the task. Where a mixer wagon is used, the person should understand the sequence of feeds to be loaded, the length of time and speed that the wagon should run, and the maximum and minimum load sizes at which the wagon can operate.
• Mix the straights prior to loading and/or purchase the feed as a mix.
• Print the diet sheets in an easily readable manner, with multiple forage DM increments for changing forage DM and animal numbers.
• Make sure scale displays are easily visible from the loading tractor. (Accurate weigh scales are a must and should be calibrated regularly.)

If the cow is able to select from within the ration (termed sorting), she may consume a very inconsistent diet. She will tend to select against long particles – greater than a palm's width – which will result in her eating meals that have greater grain content than intended, thereby leading to ruminal acidosis. Signs of sorting include:

• 'Holes' eaten into the feed, and the feed that has been rejected contains more forage and less grain than the original ration.
• Feed in the trough that looks different throughout the day, and would be different if analysed.
• A variation in the consistency, particle size, and grain content of dung.

Sorting can be minimized by:

• Avoiding excessive amounts of long material in the TMR: the longest chop should be a palm's width across.
• Adding wet feeds such as wet brewers' grains (wetter rations help the various feeds to stick together, thus making it more difficult to sort).
• Using palatable feeds, as they are less likely to be sorted than unpalatable feeds. The addition of molasses reduces sorting, particularly when added to the TMR.

## *Feed Presentation*

Prior to supply of new feed, the feed trough should be cleaned out to prevent spoilage and to remove materials that cows do not eat. Fresh feed stimulates the cow to eat, but provided the feed presented is stable and remains fresh (a function of feed constituents and environmental conditions), then feeding more than once a day will not result in an increase in feed intake. Cows like to push feed around with their noses and will eat more if they do not have to lick it off the floor, so when little is left (less than 3cm across the bottom), they will eat 50 per cent more of the remainder if it is pushed into piles.

As discussed earlier, it is undesirable that cows are shut away from feed for prolonged periods because it may result in slug feeding and reduced feed intake.

Measured feed intake for the group should be within approximately 5 per cent of that calculated in the formulated diet. There should always be feed left immediately before the allocation of the next feed and this quantity should be no less than 2 per cent of the total.

Feed should be fed along the full length of the available feed space or the effective trough space is reduced.

## Checking it Works

Given the potential for error in feed analysis, subsequent formulation, feeding the ration, and then ensuring that cows eat it, it is vital to monitor production levels and cow health and fertility to assess how the ration is performing.

### *Monitoring the Records*

#### DAIRY HERDS

*Yield* The average milk yield per day gives a general impression of the milk production level of the herd. However, the average yield is influenced by days in milk, season of calving, and the number of calves each cow has produced (parity) as well as quality of nutrition. Therefore when monitoring yield data it is important, where possible, to use prediction data, such as those provided by milk recording organizations, based on previous lactation curves for that farm, month of calving and previous lactations from the individual animal. It is also useful to assess the percentage decline in yield per month after peak. After peak heifers drop about 0.2 per cent milk per day, and older cows drop about 0.3 per cent milk per day (or 3 per cent every ten days).

*Milk quality* Monitoring milk constituents at herd or individual level can also give an indication of nutritional status. Low protein often reflects poor energy status, whereas depressed butterfat can reflect acidosis or a lack of fibre. The level of milk urea should also be noted. It is difficult to interpret, but excessively low or high levels should be investigated.

*Disease levels* Analysis of the level of periparturient disease (specifically milk fever, ketosis and displaced abomasum) should be monitored. These diseases should largely be avoidable and therefore their occurrence warrants investigation. Prevention of milk fever is relatively easily achievable, while the other diseases are more complex.

Monitoring of fertility is also a key area of assessing herd performance. Where fertility is sub-optimal, analysis of pregnancy rates and other data helps to determine the likely causes. Changes in pregnancy rate in relation to days post calving and falls in pregnancy rate with changes in nutrition suggest a nutritional component to the problem.

#### BEEF FATTENERS

*Slaughter weight and quality of the carcass* This determines the overall value of the carcass and the economic return for the farmer. A number of factors affect it, but nutrition is most important (*see also* Chapter 2).

*Stocking rate* Profitability per hectare is an essential component of farm economics and reflects ability to manage the land, particularly the grass, to best effect. Many factors influence optimal stocking rate (*see* Chapter 6),

| Condition score | Tail-head | Loin | Ribs |
|---|---|---|---|
| 1 | No fat and deep cavity | Individual transverse (horizontal) and vertical processes prominent. Horizontal processes sharp | Individual ribs obvious and sharp |
| 2 | Shallow cavity. Pin bones prominent. Some fat | Horizontal processes prominent but rounded | Individual ribs identifiable but look rounded rather than sharp |
| 3 | Fat cover over whole area. Pelvis can be palpated | Horizontal processes can only be felt with pressure | Individual ribs not defined but can be felt with pressure |
| 4 | Tail-head completely filled and folds of fat visible | Cannot palpate horizontal processes. Very rounded appearance to loin | Folds of fat over ribs developing |
| 5 | Tail and head buried in fatty tissue | Pelvis cannot be palpated | Covered in fat |

**Table 14: Physical characteristics of different condition scores**

which is a balance between maximizing beef production and managing land in a sustainable way.

*Mortality* This may or may not be related to nutrition, though many diseases are indirectly linked to nutritional status and others (e.g. bloat) are directly caused by feeding.

SUCKLERS

*The number and weight of calves weaned per cow* Many factors, including nutrition, affect the number of calves born (*see also* Chapter 7). The weight of calves is also dependent upon several factors (*see* Chapter 2), of which nutrition is the most important.

*Calving period* The calving period should be kept as close to 365 days as possible to maintain a tight calving block (*see* Chapter 7). Nutrition is one of the factors that determine how quickly cows get in calf at the beginning of the service period.

*Monitoring the Animals*

CONDITION SCORING

There are three main areas of the cow's body that can be used to score her condition: the tail-head and pelvic area, the loin, and the ribs (*see* annotated photograph on page 98). The relative condition score can then be ascribed by using the guide in Table 14.

Cow condition is a good indicator of the energy status of the animal. The loss of condition in early lactation is well correlated with the ability of the cow to become pregnant; the greater the loss the lower the pregnancy rate. Similarly there are well-accepted targets for cow condition at calving, with both excess and poor condition detrimental to the future productivity of the cow. Condition scoring is a valuable tool that is under-utilized by many farmers. The key groups to monitor are cows:

- At drying off.
- At calving.

**KEY:**

—— Tail-head and pelvis. This is assessed by:
1. The relative filling of the tail-head cavity (B).
2. The relative prominence of the pin bones (A) and hook bones (C), and the tissue cover between these points.

—— Loin. This is assessed by:
1. The relative prominence of the vertical (D) and transverse (E) processes.
2. The degree of tissue covering between these two points.

—— Ribs. These are assessed by their relative prominence and tissue covering

A. Pin bones.
B. Tail-head cavity.
C. Hook bone.
D. Vertical spinous processes.
E. Transverse spinous processes.

*Areas to be assessed when measuring condition score.*

• In early lactation (during service and pregnancy diagnosis).
• At mid-lactation.

The latter group is the most critical as the farmer can most effectively influence cow condition at this stage. Systems can either monitor the individual sentinel cow's progress throughout the lactation or use averages of groups at different stages of lactation to derive the flow of condition change.

The nutrition of the suckler cow is about management of her condition score. Table 15 (*see* page 100) shows target condition scores for both spring and autumn systems. The critical period for assessing and addressing condition score is mid-pregnancy. Condition score must be right at calving: too thin, and poor pregnancy rates and poor milk production will follow calving; too fat and there will

pin bones sharp,
with little fat

very little fat cover
around tail-head

*Dairy cow in poor
condition. She is
unlikely to get back in
calf and milk yield
may be suboptimal.*

pin bones covered in
thick layer of fat

tail-head buried in
fat and surrounded
by fat

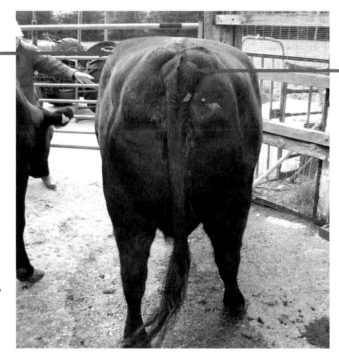

*Beef cow in overfat
condition. She is likely
to have difficulty at
calving and to lose
weight too rapidly
after she calves.*

be increased calving problems. Spring-calving cows should calve at a lower condition score because they will be on spring grass – and therefore a rising plane of nutrition – while autumn calvers need to be at a relatively high condition score at service, after which they will be managed to reduce condition over the expensive winter feeding period.

### Table 15: Optimum condition scores for suckler production

|  | Calving | Service | Mid-pregnancy | Weaning |
|---|---|---|---|---|
| Spring calving | 2.5 | 2 | 2.5 | 3 |
| Autumn calving | 3 | 2.5 | 2 | 2.5 |

### Table 16: Target body condition scores for dairy cows

| Stage | Ideal score | Range |
|---|---|---|
| Dry off | 3.0 | 3.0–3.5 |
| Calving | 3.0 | 3.0–3.5 (heifers 2.75–3.25) |
| Early lactation | 2.5 | 2.50–3.0 |
| Mid-lactation | 2.75 | 2.5 – 3.0 |
| Late lactation | 3.00 | 2.75–3.25 |
| Growing heifers | 3.00 | 2.75–3.25 |

Cow condition includes the overall impression of the cow; for example, rough stary coats suggest that cows are under some form of stress, which requires further investigation.

BLOOD

The analysis of a variety of metabolites in the cow's blood (or in certain cases milk) can provide a useful insight into the likely metabolic status of that animal and, if a sufficient number of cows are tested, the group as a whole.

FAECES

Excessively loose, hard or very variable faeces may indicate a problem with the ration or its delivery.

*Dung of normal consistency forms a pat with a peak in the middle.*

*Loose dung suggests a digestive upset and/or rumen acidosis.*

*Group of even-sized fattening beef cattle that are nearly 'finished' and ready for slaughter.*

CUD RATES

Assessing the amount of time spent cudding can provide useful information, particularly in relation to the potential for sub-clinical acidosis. The target is for at least 70 per cent – ideally 80 per cent – of a cow herd to be cudding when they are at rest and awake. An individual should chew a cud sixty times.

DRY MATTER INTAKE

Frequently the predicted intakes are different to measured intakes. In order to measure true intake of feed:

- Monitor the dry matter and volume of silage used.
- Ensure that weigh scales on mixer wagons are calibrated.

- Check that the ration on paper is the ration being fed.
- Monitor the level of wastage.

LIVEWEIGHT GAIN

Fattening stock and growing heifers should be weighed and their condition assessed regularly to ensure they are picked out for slaughter at the optimal time (*see* Chapter 2).

*Monitoring the Feed*

DAIRY

Feed conversion efficiency is measured as kilograms of milk per kilogram of dry matter intake. High-producing groups may attain values of 1.7 to 1.8. Herds with heat stress, poorly balanced rations, ruminal acidosis, long

101

## Table 17: Nutritional composition of common feeds

(Figures quoted for the forages are based on average figures. The energy, protein and NDF figures quoted are values for the dry matter of the feed.)

| | Dry matter % | Energy (MJ / ME) | Protein % | NDF % |
|---|---|---|---|---|
| **Forages** | | | | |
| Grass | 20 | 11.2 | 15.6 | 57.6 |
| Hay | 87 | 8.8 | 10.7 | 65.7 |
| Grass silage | 26 | 10.9 | 16.8 | 58.2 |
| Dry lucerne | 90 | 8.6 | 19.2 | 47.3 |
| Lucerne silage | 34 | 8 | 19.4 | 49.5 |
| Maize silage | 30 | 11.5 | 8.8 | 39 |
| Spring barley straw | 86 | 6.8 | 4.4 | 80.5 |
| Urea wholecrop wheat | 55 | 10.4 | 23 | 50 |
| Kale | 13 | 11.8 | 16.4 | 25.7 |
| Rape | 14 | 9.5 | 20 | – |
| Sugar beet | 23 | 13.7 | 4.8 | – |
| | | | | |
| **High-energy concentrates** | | | | |
| Barley | 87 | 13.3 | 12.9 | 20.1 |
| Wheat | 87 | 13.6 | 12.3 | 16.6 |
| Caustic wheat | 77 | 12.6 | 11.8 | 11.4 |
| Citrus pulp | 89 | 12.6 | 7.2 | 22.8 |
| Fodder beet | 18 | 11.9 | 6.3 | 13.6 |
| Maize gluten feed | 89 | 12.7 | 23.3 | 39 |
| Molasses cane | 75 | 12.7 | 4.1 | 0 |
| Potatoes | 20 | 13.4 | 10.8 | 7.3 |
| SBF (dry molassed) | 86 | 12.5 | 12.9 | 29.4 |
| Wheat feed | 89 | 11.9 | 18.1 | 36.4 |
| | | | | |
| **High-protein concentrates** | | | | |
| Brewers grains | 28 | 11.7 | 24.5 | 57.2 |
| Cottonseed meal | 92 | 11.1 | 37.5 | 39 |
| Groundnut meal | 92 | 13.7 | 49.5 | 18 |
| Peas | 87 | 13.5 | 37.4 | 11.6 |
| Rapeseed meal | 90 | 12 | 41.8 | 27.9 |
| Soyabean meal (exp) | 90 | 13.5 | 50.4 | 29 |
| Urea | 99 | 0 | 287.5 | 0 |
| Wheat dist. grains | 90 | 12.4 | 32.2 | 33.5 |
| Whey | 43 | 14 | 24 | 10 |

*(From Chamberlain, A.T. and Wilkinson, J.M.,* Feeding the Dairy Cow, *Chalcombe Publications, 1996.)*

days in milk, and so on, may have values lower than 1.2. A sensible target figure is 1.4kg of DMI per kg of fat-corrected milk produced.

FATTENING STOCK

Feed conversion efficiency is measured as kilograms weight gain per kilogram dry matter intake. This is particularly important for finishing cattle because feed conversion efficiency falls as they near finishing weight.

Slow growers should be slaughtered as soon as they reach an acceptable fat class, because retaining them to gain a few extra kilos is not economical. In contrast, cattle that are still growing fast can be kept as long as they remain in an acceptable fat class to maximize their killing out. In other words, cattle should be kept only while the value of any weight gain is higher than the cost of creating that gain.

# CHAPTER 6

# Grass and Grazing Management

The extent to which different farms rely on grazing varies enormously, from extensive beef units – whose entire output is based on grazed, permanent pasture – to high-yielding dairy herds where grass forms only a small part of the ration. Since grazing is by far the cheapest form of forage, utilizing it to best effect should be a primary objective on all cattle farms.

Grassland management should aim to ensure a consistent supply of as good-quality feed as possible for the herd, while maintaining soil structure and preventing damage to vegetation in areas of natural or semi-natural grassland. For farmers in Europe there has been a shift in farm payments away from production and towards environmental benefits. This requires conservation management to protect biodiversity (animals, birds and plants) in the pasture and in hedgerows.

How grass grows is discussed in this chapter, before the various factors involved in successful grassland management are explored.

## GRASS GROWTH

There are two different forms of grass growth:

1. Vegetative growth ('tillering'), where the plant produces new shoots at ground level that will form new plants.

2. Reproductive growth, where most of the energy and growth is directed to the creation of the seedhead.

Tillering is predominantly exhibited in the early spring and in autumn, with reproductive growth seen in between these periods. The grazing of the plant stimulates tillering, so grazing can be used to aid in maximizing grass productivity. This is best achieved by ensuring that:

- The pastures are grazed intensively during the spring and autumn.
- There is optimum level of grazing, which allows sufficient growth to ensure tillering and reasonable yields of grass without excess seedhead production.

The level of grass growth that a sward will achieve is influenced by a number of factors:

- The management of grazing patterns.
- The land's ability to support grass growth (termed its site class).
- Temperature.
- Amount of daylight.
- Species of grass in the sward.
- Type and amount of fertilizer used.

The target is to achieve a leafy, dense sward; bare patches, brown, dead plant material and the ingress of weeds are all undesirable and will affect the quality of the grazing.

Grass growth follows a seasonal cycle. In the UK, the maximum growth peak occurs in early spring, followed by a decline in summer; there is a second small peak in the autumn.

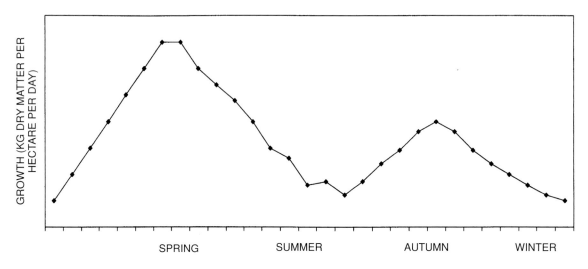

*Graph showing seasonality of grass growth.*

## GRAZING SYSTEMS

### Intensive Rotational Grazing

With this system, the total grazing area is divided into several paddocks (usually around twenty-five to thirty), which are grazed and then rested before being grazed again, in a rotational cycle. Each paddock is grazed very hard and for a short period of time – usually for only one to two days – so as to avoid excessive damage to the grass plant by overgrazing and trampling, as this will reduce the pasture quality and limit regrowth.

Intensive rotational grazing maximizes a farm's grass-production potential in terms of maintaining both the quantity and quality of grass available; as such it is the system of choice. However, it requires planning and it is also labour-intensive.

*Spring grass at the vegetative growth stage.*

| Table 18: Example of a typical level of grass growth in the UK | | | | | | | | |
|---|---|---|---|---|---|---|---|---|
| | *Feb* | *Mar* | *Apr* | *May* | *Jun* | *Jul* | *Aug* | *Sept* |
| Kg of dry matter of grass growth per hectare daily | 20 | 40 | 80 | 100 | 70 | 40 | 30 | 40 |

*Rotational, or paddock, grazing. Grassland is divided into paddocks in each of which cattle graze for one to two days before being moved on to the next.*

## Set Stocking

This system relies on matching expected requirements with expected growth and then allocating the appropriate number of stock to the appropriate area. It has the advantage of being

far simpler to operate than rotational grazing: the only changes needed are to change area supplied or number of stock according to growing conditions, and to supplement forage if grazing pressure outweighs the supply of adequate grass. The disadvantage is that it is less efficient in terms of maximizing the potential of an area to produce grass.

## ASSESSING SWARDS

The greatest challenge to the farmer when managing cows at grass is the potential differences in the level of grass consumed on any given day as a result of changes in grazing and sward conditions due to climatic and day-length variation. In high-yielding cows such variations can have serious metabolic consequences. Ensuring therefore that cows are able to maximize their potential intake from grazed grass requires:

### Stocking Rates

Generally high stocking rates provide the highest yields of grass per hectare. Stocking density can be calculated according to the required level of performance (target growth rates or milk production), breed of cattle, and ability of the land to produce grass (site class). High stocking rates cannot be achieved on natural or semi-natural grassland, which is particularly sensitive to overgrazing. Farm layout, specifically access to fields, is important for the dairy herd which must be able to return to the farm twice daily for milking.

*Stocking density must be low on moorland grazing to preserve the pasture.*

- An ability to assess the quantity and quality of the grazing available.
- An ability to adjust the grazing allowances on the basis of these assessments – i.e., not just following a target stocking density.

Consequently, pastures (and when appropriate animal performance) should be evaluated on a regular basis to allow sensible decisions, both for the present and the future requirements, to be made in the management of the grazing. There are various methods that can be employed to assess the quantity and quality of grass growth.

## Height of the Grass

The suitability of a pasture for grazing is traditionally assessed by evaluating the height of the grass (using the 'welly boot' test); Table 19 shows optimal figures for the season and grazing system. While grass height is a good basic indicator, it does not take account of the density of the sward. This is important because for any given height of grass, the density (and consequently the amount of dry matter) may vary enormously. This is particularly true in swards containing high levels of clover, which will be much denser than in a pure grass ley.

## Sward Density

This can be assessed by eye (though this requires experience), or more objectively using a plate meter. This meter consists of a plate which fits over a stem or rod held in a vertical position with its base at ground level. When the plate is allowed to rest on the sward, the instrument's readings are influenced by a combination of sward height and density.

## Quality

In order to ensure paddocks are grazed at the correct time and for an appropriate duration it is important to be able to assess the level of growth the grass is showing. The best method of doing this is to inspect a number of individual grasses within the paddock and count the number of leaves the plant has produced. This will provide an assessment of the overall sward's productivity and maturity and is based on the principle that a ryegrass tiller typically supports three green leaves.

| Table 19: Target sward height (cm) for rotational and continuous grazing | | | | |
|---|---|---|---|---|
| | *Spring* | *Summer* | *Autumn* | *Winter* |
| Rotational | 8 | 10 | 12 | <5 |
| Continuous | 6 | 8 | 10 | <5 |

*The 'welly boot' test to assess sward height.*

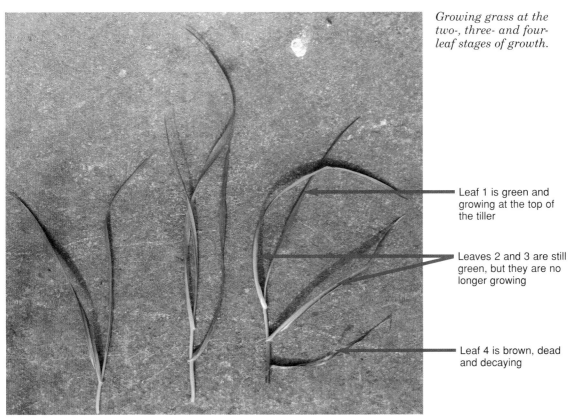

*Growing grass at the two-, three- and four-leaf stages of growth.*

Leaf 1 is green and growing at the top of the tiller

Leaves 2 and 3 are still green, but they are no longer growing

Leaf 4 is brown, dead and decaying

Two leaves: too early to graze

Three leaves: ideal for grazing

Four leaves: too late for grazing

107

*The nutritional requirements of spring-calving suckler cows are easily met by grazing on permanent pasture.*

If the ryegrass is grazed when it has fewer than two fresh (not remnant) leaves, the vigour of the plant and subsequent regrowth is reduced. Conversely, allowing cows to graze ryegrass that has more than three leaves reduces sward productivity because leaf death and decay have already commenced and the lower leaves may be shaded, thereby reducing the density of the overall regrowth. In the UK there is an exception to this rule during the periods of peak growth seen in May and June. During this period, waiting for the emergence of the third leaf will mean that the grass will have grown too far (owing to the speed of growth), and consequently the grass will be more suitable for mowing than for grazing.

## PLANNING GRAZING

### Set Stocking

Basic calculations for the requirement for grazing area during the year can be made knowing the likely grass growth rates on the farm. Because many different factors influence growth rates, every farm varies in its ability

to support grass growth. It is better to develop an understanding of the individual farm and plot its growth rates yourself than to rely on area or national growth-rate statistics. Once these rates have been assessed, the requirement calculations can be made as follows:

Calculate expected required daily dry matter intake from grass. This figure will depend on a variety of factors, including:

- Nutritional requirements. As the quality and density of grass falls during the summer, the energy density of the grass may become too low to support higher-yielding dairy cows, and be suitable only for low-yielding or dry cows. Grass quality later in the season may also be too low to support target growth rate in beef cattle, and they should be given supplements or brought inside to finish.
- Dairy cow production level. In a high-yielding herd, there must not be too many fluctuations in dry matter intake as a result of variable grazing conditions; cows

should be buffer-fed to a safe margin, thereby reducing the intake requirement from grass.

- Availability of other forages. There may be alternatives to grass – such as kale or stubble turnips – available on the farm.

The total yield of grass required (not including silage to be cut) can then be calculated by multiplying daily intake by the number of stock. This figure can then be divided by the average figure for growth per hectare per day to determine the grazing allocation for that period of the year.

Take, as an example, a 100-cow dairy herd. Assume a required dry matter intake from grass to be 12kg for each individual. With 100 cows this would mean a daily requirement of 1,200kg dry matter of grass. In May the growth rate is 100kg of dry matter per hectare per day; therefore 12 hectares (1200/100) would be sufficient to support this number of cows. In August, however, the growth rate is only 30kg of dry matter per hectare per day, so the area required would be 1200/30 = 40 hectares.

This figure leaves no margin for error if growth rates are lower, and cattle will not be 100 per cent efficient in their grazing (no more than 75 per cent of grass dry matter grown is likely to be consumed), so generally a greater area than this should be offered. The extra buffer area can be fenced off and fed only if required; if it proves not to be required it can be cut for conservation.

It can be seen from these figures that only a small proportion of the total grazing is required in the spring; the remainder will be cut for silage. As grazing area requirements increase as the season progresses the cattle can graze the aftermath, or it can be fertilized and cut again later in the season. If the area required matches or exceeds the total area available for grazing then supplementation will be required to ensure the cows' dry matter requirements are met.

## Rotational Grazing

The aim of a rotation is to enable cows to come to a new allocation when it has reached its optimum pre-graze target (*see* Table 20), and then to remove the cows before the field goes over the post-graze target. The first field is regrazed once the other available fields have been grazed; the trick is to ensure that each field reaches its pre-graze target at exactly the right time. The key to achieving this outcome is to adjust the length of the rotation between grazings to accommodate changing growth rates, either by taking land out or by increasing the area to be grazed.

To calculate the requirements for rotation area and frequency, two steps need to be taken.

1. Calculate the area required (assuming a target of twenty-four hours per paddock):

Daily dry matter intake of grass × number of cows

divided by

Difference between start and end grazing target for that period of the year

---

### Table 20: Target amount of grass cover per hectare (kg/DM)

|  | *Start target* | *End target* | *Difference* |
|---|---|---|---|
| April | 2500 | 1500 | 1000 |
| May–August | 2800 | 1500 | 1300 |
| September | 3500 | 1500 | 2000 |
| October–late November | 2500 | 1500 | 1000 |

*(Source: Milk Development Council. Grass and Grassland Management Improvement Scheme.)*

In the example 100-cow herd eating 12kg of DM per day in May and July the calculation is:

$$1200\text{kgDM} \times 100 / 1300\text{kgDM} = 0.9 \text{ hectare}$$

2. Calculate the time for the paddock to return to pre-grazing target:

Difference between start and end grazing target for that period of the year

divided by

Growth rate for that period of the year

For May the calculation is
1,300kgDM / 100kgDM = 13 days

For July the calculation is:
1,300kgDM / 40kgDM = 33 days

As growth slows down, the rotation length has to be extended by bringing more pasture into the grazing cycle. If grass growth slows down too far – i.e., below 40kg per hectare per day – then it will be necessary to supplement if target dry-matter intakes are to be met while at the same time avoiding excessive numbers of paddock changes.

# PRACTICAL CONSIDERATIONS

## Fencing

To be able to alter the grazing in response to the changes in sward condition and growth that we have discussed, electric fencing is going to be required. This will allow for the amount of pasture allocated to the cattle to be increased or decreased, depending on the level of grass growth, with excess areas kept aside for cutting if they are after all not required. This system is termed 'buffer grazing'.

Where a field is sub-divided into paddocks it is important to ensure that each paddock has a water supply and that back-fencing is used to prevent cattle getting access to recently grazed ground. This will protect pasture regrowth by avoiding excess defoliation and also prevent compaction and poaching of the ground.

*Temporary electric fencing enables you to adjust the paddock area available to the animals: this allows you to match the animals' feed requirements with the level of grass growth.*

*Good-quality cow tracks protect grazing from trampling and reduce the risk of lameness.*

## Field Access

If cows are to access grazing quickly and efficiently, without injuring themselves or damaging the grazing, then suitable cow tracks will be required. All too frequently cows are rendered lame through walking through badly maintained and designed gateways and tracks. While such tracks require a large amount of capital investment, provided they are designed correctly the improvement to cow health and mobility will rapidly repay the initial cost. If cows have to walk through one field to get to another, temporary access should be provided, with an electric fence parallel to the existing hedge or fence to minimize the damage to the ungrazed paddock.

## Turnout

At turnout the first pastures to be grazed should be close to the farm, to allow close supervision of the stock, and have good access and good drainage to minimize poaching. The transition to grazing should be gradual – a period of two to three weeks or longer is desirable – as this allows for a gradual change in diet. Initially cows should be allowed access to pasture for only a few hours per day. They should remain housed overnight so that control over nutritional balance can be maintained. It is important to emphasize that dry matter intake is the critical factor during this period: intakes should not be compromised in pursuit of grazing. Practically, this means that many cows require buffer feeding during this period, and they should be brought in early for milking to allow adequate time for this. One advantage of buffer feeding is that it allows a better determination of grass dry matter intakes (by subtracting buffer eaten from expected total dry matter intake).

## Dry Cows

Particular care must be taken with the feeding of transition cows (close to calving) on grass. Grass is high in calcium, potassium and sodium, all of which increase the risk of milk fever. Ideally these cows should be housed, but certainly grass intake should be controlled. Dry cows at pasture should graze unfertilized, permanent pasture rather than heavily fertilized new leys.

## Minerals

Grass frequently does not meet the mineral and vitamin requirements of cattle, so supplementation of minerals is often required (the exact amount dependent on the area). Grass 'staggers' is a fatal condition resulting from a lack of magnesium, seen particularly in the spring and autumn, when levels of this mineral in grass are especially low. Magnesium should always be on offer during risk periods. Various methods are available for supplementing minerals and vitamins, and these should be discussed with a veterinarian.

## Shortfalls and Surpluses

Where assessment of sward growth indicates that there is, or will be, an excess or deficiency of grass, this should be managed. In the case of a surplus, short-term solutions include topping or cutting for conservation; longer-term remedies are to increase stocking density or reduce fertilizer inputs. In the case of a shortfall, short-term solutions are to increase grazing area or the level of supplementation and, in the longer term, to reduce stocking density or increase fertilizer use and improve the ley quality by re-seeding.

# ENVIRONMENTAL BENEFITS

Good grassland management should not only supply the nutritional requirements of stock but it should also benefit the environment. It can improve soil structure because grazing pasture increases organic matter in the soil, and the network of fine grass roots stabilizes the topsoil.

However, overgrazing, or grazing land that is too wet, can result in erosion and damage to soil structure, which in turn cause poor grass growth rates and poor land drainage. The risks can be minimized if:

- Cattle are not grazed when land is too wet. In the UK this usually means that suitable housing should be available.
- Heavy machinery is not used on wet land. This is a particular problem with spreading slurry or manure in wet conditions and during the winter when jobs such as hedge-trimming and fencing are carried out.
- Well-drained tracks are installed so vehicles do not have to use pasture.
- Supplementary feeders, if used, are moved regularly and placed on freely draining soil.
- Re-seeding is carried out early enough in autumn to achieve good ground cover before winter.

It is important to maintain biodiversity on all grazed land, but there are particular risks with natural and semi-natural grazing on upland, hills, moors, and marshy areas. Loss of ground cover is likely to lead to erosion and should be avoided. Stocking densities must be low enough to avoid overgrazing, vehicle use should be kept to a minimum, and supplementary feeders should be sited away from sensitive vegetation. Damage to riverbanks can be kept to a minimum by providing stock with alternative sources of water.

Fertilizers tend to alter the natural vegetation in permanent pasture: they favour grasses over other species, so, with the exception of rotted manure, they cannot be used in conservation areas. Forage cannot be cut until late (at least one year in three) and must be left lying for at least thirty-six hours to allow time for the seeds to fall into the ground.

Hedges, which for many years were removed to increase field size, are to be preserved. They can only be trimmed in winter (to avoid disturbance to nesting birds and budding hedgerow plants).

# SUPPLEMENTING GRAZING

Given that grazed grass is by far the cheapest feed available to the farm, if supplementation is deemed necessary the priority should be maximize its effect on increasing production while at the same time minimizing the effect it has on reducing grass intake. Supplementary feeds will be required if it is considered that the grazing on offer does not meet the required forage dry matter intakes or nutritional requirements for the animals concerned. The most common nutritional imbalances and deficiencies in dairy cattle include:

1. An excessive amount of degradable protein and an insufficient amount of fermentable energy in the grass.
   This means the rumen microflora cannot utilize the protein efficiently. The waste protein must be processed into urea by the cow's liver, which costs energy and is undesirable. The consequences include:

   - Loose manure.
   - Reduced milk production.
   - Reduced milk fat.
   - Loss of body condition.
   - Increased excretion of nitrogen.

   This relative imbalance can be counteracted by providing a supplement high in fermentable energy.

2. Insufficient undegradable protein for high-yielding dairy cows.
   Such cows will need direct supplementation from an appropriate protein source.

3. Low fibre content in high-quality pasture (particularly in the spring and autumn).
   Stimulation of cud chewing and rumination will be reduced and may result in reduced milk fat content.

4. Low levels of dry matter.
   A fall in sward DM from 20 per cent to 15 per cent means cows need to consume an additional 25kg of fresh sward to maintain the same daily DM intake. In practice, the lower the sward DM, the more likely it is that cows will be unable to achieve target DM intakes because they run out of grazing time. Cows are unlikely to eat any more than 70kg of grass as fresh weight.

Fattening cattle require supplementation when the grass nutritional quality falls to a level at which it no longer supports target growth rates. This usually occurs from mid-summer onwards.

## When to Supplement

The effect of supplementation on herbage intake depends on availability of grazing, the type of supplement provided, and the breed of animal being supplemented. When grass availability is low, concentrate supplementation will hardly affect grass intake; when grass availability is high, the substitution rate is high. In other words, when high-quality grass is readily available, supplementary feeding will always reduce grass intakes.

In general, forage supplements have high substitution rates whereas concentrate supplementation has less of an effect. Although conserved forage supplementation may improve total DMI, it should not be used if grazing supply is adequate. The exception is where cows are unable or unwilling to eat sufficient grass during grazing, specifically where:

- There is inadequate grazing (e.g. average farm cover estimates are below target and less than 1,800kg DM per hectare).
- There is poor-quality grazing (wet, low energy and protein content, or low DM content).
- There is a high proportion of cows yielding over 25 litres or there are heifers in the herd. The higher yielding the animal, the more important it is that dry matter intakes are always met to prevent metabolic compromise. Higher-yielding cows will need a greater safety margin in relation to the amount of grass they can be expected to eat

*Supplementary feeding at grass may cause severe poaching. This field will take months to recover and to start producing good grass growth again.*

on a given day. Consequently these cows will require more buffer for any given level of sward cover.

## Practical Aspects of Supplementation

Buffer feeds for dairy cows should be given in the early afternoon to maximize evening grazing and consequent grass DMI, although most supplementation is given post evening milking. If supplementation is to take place in the field, as with beef cattle, then care must be taken to avoid excess poaching of the land around the feeders. This can be done by ensuring that feeders are placed on level ground, well away from water courses. Supplementary feeders should be moved regularly and, where possible, tracks should be available for vehicles.

## Grazing High-Yielding Cows

High-yielding cows have the potential to produce more milk from grazing than less productive animals because their appetite is higher, but feed intake is the major limiting factor of grazing high-yielding cows. Cows at pasture, even when grazing high-quality pasture, typically consume 20 per cent less dry matter than housed cows. Under optimal con-

ditions, lactating cows (Holsteins) can be expected to consume around 17–18kg DM per cow per day. This amount of intake may support 23–25 litres of milk in early lactation based on the estimated energy intake.

The critical factors in ensuring success are:

- Achieving a high sward quality. The most important factor governing intake is the amount of feed taken in per bite, so sward quality needs to be high to allow each bite to be maximally effective.
- Appropriate supplementation to ensure that intakes are maximized and to match any nutritional deficiencies within the grazing.

In this chapter we have considered how the productivity of grassland can be maximized for profitable production. In some regions, particularly hill areas and marshlands, the emphasis is upon sustainability and maintenance of a diverse sward that includes a range of plant types, and this is reflected in current farm payments in the UK, where positive environmental outcomes are required under the single farm payment scheme. These issues are beyond the scope of this book, but may dramatically change a farm's approach to management of grassland.

# CHAPTER 7

# Management of Fertility

The fertility of the herd is central to its production success, and good fertility management is an important facet of cattle keeping. Cattle fertility is influenced by many factors, particularly nutrition and disease. Any imbalance between the animal's needs and the quality or quantity of feed eaten is likely to result in reduced reproductive performance. Disease, even if it has no direct effect on the reproductive process, is also likely to reduce fertility – often by reducing appetite and feed intake.

This chapter examines the range of issues that determine reproductive success and hence the performance of the herd. The chapter is divided into the following sections:

1. The anatomy and physiology of the cow.
2. Fertility management.
3. Management of the bull.
4. The calving cow.
5. Principles of genetic selection.

## THE ANATOMY AND PHYSIOLOGY OF THE COW

The anatomy of the cow's reproductive tract is outlined in the diagram on page 116.

### Puberty

The age of puberty (onset of ovarian cyclicity) in the heifer varies, generally occurring at seven to eight months, when the heifer reaches 40 to 45 per cent of her mature body size. The range is four to fourteen months, the variation dependent on the animal's breed, growth rate and body condition.

### The Reproductive Cycle

The reproductive cycle of the cow has a range of eighteen to twenty-four days but on average lasts for twenty-one days. Every cycle begins with the process of ovulation: the release of a mature egg (ovum) from the ovary. The egg passes into the Fallopian tube where, if it meets fertile sperm, it may be fertilized. Once fertilized, it moves into the uterus where it develops, implanted in the uterine wall to derive its nutrient requirements from the mother. If all goes well, the calf is born 283 days later. If the ovum is not fertilized, or if the early embryo is lost, the cow will produce another ripe egg and cycle again after twenty-one days.

The reproductive process is controlled by the ovarian hormones (oestrogen and progesterone), and the pituitary hormones – follicle stimulating hormone (FSH) and luteinizing hormone (LH). Under the influence primarily of FSH the ovum matures in a fluid sac or follicle, which increases in size producing increasing amounts of oestrogen as it grows. This increase in oestrogen stimulates the cow to show signs of oestrus or, more colloquially, 'bulling' or 'heat', during which time (approximately twenty-four hours within a range of six to thirty-six hours), she is receptive to the bull and can be served or inseminated. The increase in oestrogen also causes a surge in the release of the hormone LH, which in turn causes the

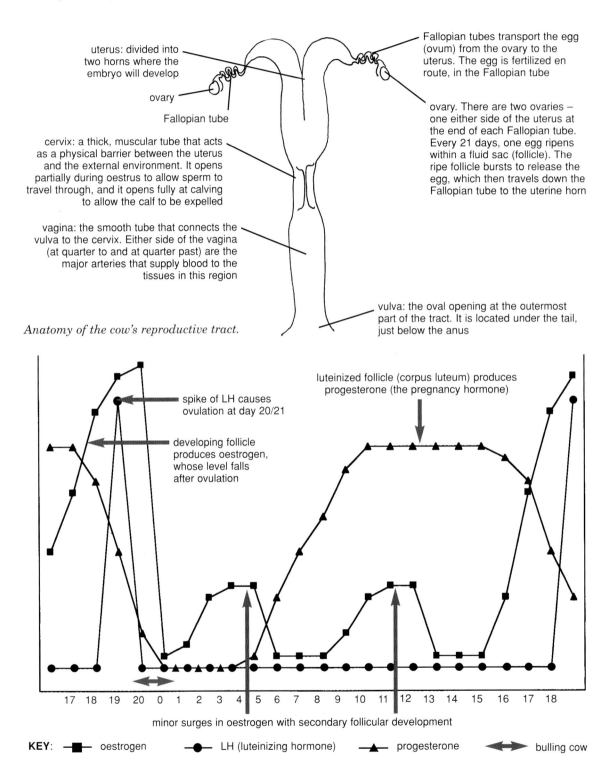

uterus: divided into two horns where the embryo will develop

ovary

Fallopian tube

Fallopian tubes transport the egg (ovum) from the ovary to the uterus. The egg is fertilized en route, in the Fallopian tube

ovary. There are two ovaries – one either side of the uterus at the end of each Fallopian tube. Every 21 days, one egg ripens within a fluid sac (follicle). The ripe follicle bursts to release the egg, which then travels down the Fallopian tube to the uterine horn

cervix: a thick, muscular tube that acts as a physical barrier between the uterus and the external environment. It opens partially during oestrus to allow sperm to travel through, and it opens fully at calving to allow the calf to be expelled

vagina: the smooth tube that connects the vulva to the cervix. Either side of the vagina (at quarter to and at quarter past) are the major arteries that supply blood to the tissues in this region

vulva: the oval opening at the outermost part of the tract. It is located under the tail, just below the anus

*Anatomy of the cow's reproductive tract.*

spike of LH causes ovulation at day 20/21

developing follicle produces oestrogen, whose level falls after ovulation

luteinized follicle (corpus luteum) produces progesterone (the pregnancy hormone)

minor surges in oestrogen with secondary follicular development

KEY: ■ oestrogen   ● LH (luteinizing hormone)   ▲ progesterone   ⬌ bulling cow

*Graph showing the hormonal changes that occur during the cow's reproductive cycle. The cycle is repeated if the cow does not become pregnant.*

## Signs of Oestrus

The first thing that the stockperson may notice is some changes in behaviour. A cow that normally comes into the milking parlour early may not appear until much later. She may not finish all her concentrate and may be restless.

The so-called 'bulling string' is clear mucus or slime. It may be hanging from the cow's vulva or around the tail or it may be on the back of the cubicle when she stands up. The mucus changes to blood after oestrus – at this point it is too late to serve the cow.

Bulling cows may become restless and noisy. These signs are more obvious when there are several animals bulling at the same time. The bulling cow may butt other animals and she may become more vocal, bellowing or mooing at others in the group.

The bulling cow may sniff at the vulva of other cows – particularly any animal that is also coming bulling – or sniff at urine of other cows.

Other cows may attempt to mount a bulling cow; at this stage she is likely to move away.

The bulling cow may rest her chin on other cows. This is often a preliminary to trying to mount, or she may stand resting her chin.

She may mount or attempt to mount other cows.

Sometimes she may 'head mount' other cows, mounting on the head or shoulder of other cows.

She may also stand to be mounted. This is the most valuable sign to show that a cow is bulling: if a cow stands to be mounted without attempting to move off it is certain that she is bulling.

INCREASING VALUE AS SIGNS OF OESTRUS

follicle to ovulate and to develop into a structure called the corpus luteum (CL). After ovulation, the levels of oestrogen fall.

Once developed the corpus luteum (CL) releases progesterone, the hormone responsible for maintaining pregnancy. If the cow conceives, the CL continues to produce progesterone throughout pregnancy. The presence of the CL partially inhibits release of the hormone FSH so no further reproductive cycles occur. (While the CL is present, development of new follicles does occur, but the presence of progesterone prevents them from progressing into mature follicles capable of ovulating. These follicles will regress. There will be one or two 'waves' of this unapparent follicular development and regression during the life of a CL.) In the absence of a viable embryo the CL has a natural 'shelf-life' of approximately fourteen days before a series of hormonal events results in its destruction. The disappearance of the CL and consequent fall in progesterone removes the inhibition of FSH, which is released to stimulate a new follicle to develop, with release of oestrogen.

The diagram opposite (bottom) illustrates the hormonal events associated with the cow's oetrous cycle.

## Signs of Oestrus

During oestrus the cow demonstrates some specific behavioural and physiological changes,

*Cow sniffs at the vulva of a bulling animal.*

*The 'bulling string'.*

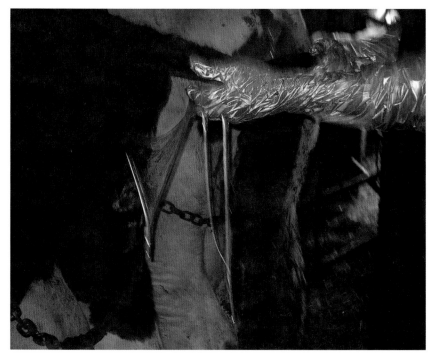

*Standing to be mounted is the best sign of bulling.*

which, if artificial insemination is to be used, must be observed by the stockperson so that insemination can be correctly timed to coincide with ovulation.

The most fundamental and only truly accurate signal that the cow is in oestrus is that she will stand to be mounted by other cows. Other behavioural and physical signs, in order of accuracy as indicators that the cow is in oestrus, are detailed on page 117.

## POST-SERVICE

There are three potential outcomes after a cow in oestrus is served or inseminated:

1. The cow conceives and maintains the pregnancy.

2. The cow fails to conceive.

3. The cow conceives but the pregnancy fails shortly afterwards. This is termed early embryonic death and is a frequent occur-

rence. It is estimated that around 30 to 40 per cent of pregnancies fail within the first thirty days and a further 15 per cent by 60 days. If the pregnancy fails within the first two weeks, the cow returns to oestrus within the normal period following the previous oestrus, and there is no indication that the cow conceived at all. If the pregnancy fails after fifteen to sixteen days then there is likely to be a delayed return to oestrus beyond the maximum of twenty-four days. Early embryo death may be caused by genetic, nutritional or infectious factors.

If a pregnancy is successfully established, the developing embryo releases a signal to the cow preventing the destruction of the CL. Progesterone becomes the dominant hormone and remains so through the pregnancy. This in turn means that the cow no longer produces mature follicles or shows signs of oestrus (though approximately 5 per cent of cows continue to show some signs of oestrus behaviour during pregnancy).

119

## ABORTION

Not all cows manage to maintain a pregnancy through to delivery of the calf, and an abortion rate of up to 4 per cent annually should not be considered abnormal. Rates above this should trigger an investigation into the possible cause. Abortions may be caused by a variety of factors, not of all them infections in origin. There are specific infectious causes of abortion, and these can cause severe welfare and economic problems. In the UK, all abortions should be reported to DEFRA.

## POST-CALVING RETURN TO OESTROUS CYCLING

If cows are to maintain a calving pattern of 365 days, they must rapidly resume normal oestrous cycles with a uterus that is clean and capable of supporting another pregnancy.

The uterus takes around forty days to contract fully and become free from infection after calving. In dairy cows, provided they are not in severe energy deficit, oestrous activity will normally resume around thirty days post calving. In suckler cows, this process is inhibited by suckling so that only 90 per cent of normal suckler cows have started to cycle again by eighty days after calving. Trauma, the retention of the placenta, or the development of uterine infections, delay or even halt this process, rendering the cow infertile or sick until the condition is treated. The susceptibility of the cow to such problems is affected by her metabolic status. Metabolic stress or disease also delays her return to normal oestrus cycling.

In other words, disease or nutritional stress during the transition period around calving increases the likelihood of reproductive disorder. The cow's management at this time must be optimal. Any cow showing problems should be assessed and treated appropriately. The major problems are:

• Retained placenta.
• Metritis.
• Endometritis.

• Abnormal ovarian cyclicity, or failure to return to normal ovarian cyclicity.

### Retained Placenta

This occurs if the cow fails to complete the third stage of parturition and the placenta (afterbirth or 'cleansing') is held for longer than four hours post calving. The retention of the placenta will not necessarily make the cow ill; however, it increases the chance of the development of metritis, a severe infection of the uterus. Any affected animal should be monitored carefully, veterinary checks undertaken, and appropriate treatment provided.

### Metritis

Technically termed 'puerperal metritis', this condition usually accompanies, or is secondary to, retained placenta. It is characterized by a foul-smelling vaginal discharge. It invariably compromises the cow's health and is potentially life-threatening. Prompt veterinary treatment is vital.

### Endometritis

This condition (commonly termed 'whites') is caused by an infection of the inner lining of the uterus. It is characterized by a creamy discharge with little or no odour from the vulva (though not all cases of endometritis will show an obvious discharge). The cow is not sick. This condition should not necessarily be treated until twenty-five days post calving as many cows self-cure. After this time affected cows should be treated because infection renders the cow infertile and delays the return to oestrous cycling by interfering with the normal hormonal pathways.

### Abnormal Ovarian Cyclicity

Failure to return to normal ovarian cyclicity, and abnormal ovarian cyclicity, are largely a reflection of the animal's metabolic status, associated with negative energy balance and metabolic stress. Hormonal treatments may be used to treat the conditions, but underlying nutritional and disease factors should also be addressed.

*Retained placenta is commonly associated with milk fever and fatty liver. It is often followed by fertility problems.*

# FERTILITY MANAGEMENT

This includes:

- Management of the detection of oestrus.
- Management of service and pregnancy.
- Maintenance and analysis of fertility records.

## Management of Oestrus Detection

In order to calve every 365 days, the cow must become pregnant within eighty-two days post calving (assuming a pregnancy of 283 days). Where no bull is used, the ability of the farm personnel to detect that the cow is in oestrus is critical.

In dairy herds, 40 to 50 per cent of cows exhibit behavioural signs of oestrus for only six hours, often at night, and, especially in high-yielding herds, they often show few overt signs of oestrus, with only 50 per cent standing to be mounted. Therefore, oestrus detection requires a great deal of effort. To maximize the chances of success there are several areas, or factors, that should be considered:

- The personnel.
- Cow identification.
- The environment.
- Nutrition.
- Disease.
- Heat detection aids.

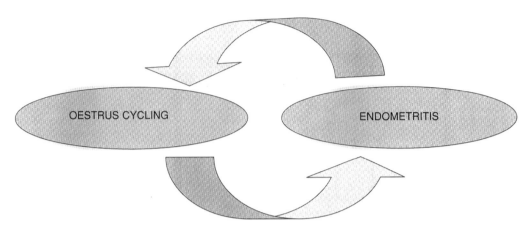

the infected uterus interferes with hormonal
status and prevents normal cycling

OESTRUS CYCLING

ENDOMETRITIS

the bulling cow is better at eliminating
infection from the uterus

*The link between uterine infection (endometritis) and the reproductive cycle.*

*Endometritis is an infection of the uterus lining that causes reduced fertility.*

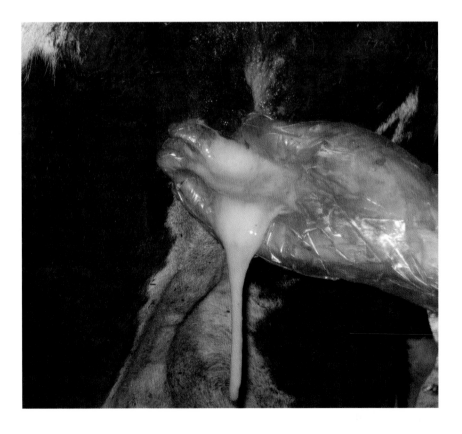

- Artificial induction and synchronization of oestrus.

## The Personnel

All members of the farm's staff should be trained to recognize and correctly identify the cow in oestrus. The stockperson should reserve at least three twenty-minute slots per twenty-four hours solely for observing cows. The best times to observe cows are away from peak activity, for example when they are at rest pre-morning milking and late in the evening.

Bulling dates should be carefully recorded. Stockpersons can then predict which animals are due to come bulling again and can observe them more closely. The information should be posted on a display board for all farm staff.

## Cow Identification

Even if a cow is observed bulling it may be difficult to identify her as a particular individual. Cows should be clearly identified by freeze branding, neck collars or eartags. Housing should provide sufficient lighting to allow clear observation and identification of the cows.

## The Environment

Cows are less likely to show overt signs of oestrus if they are overstocked with little or no loafing area or if they are standing on slippery or uneven surfaces. Where there are very large numbers of cows in one group, or where cows are moved from one group to another, the bulling cow may not be sufficiently relaxed to show normal bulling behaviour. This is a particular problem with first-calved heifers.

It can be difficult to observe cows that are grazing far from the farm, so it may be necessary to increase time spent in oestrus detection and utilize management aids (*see* heat detection aids below) to help detect oestrus.

## Nutrition

Cows in negative energy balance, as a result of insufficient energy intake for production output, are far less likely to show clear signs of oestrus. Certain mineral deficiencies have also been linked to reduced bulling activity.

Monitoring energy balance can be performed on farm by a variety of techniques (*see* Chapter 5).

## Disease

The most obvious disease factors to affect bulling behaviour are lameness and uterine infection. Cows that are sore on their feet are less likely to stand to be mounted and are more likely to be suffering from negative energy balance, and lameness inhibits the reproductive hormonal cycles. Cows with a uterine infection may fail to cycle owing to inhibition of the normal hormonal processes. Ultimately, any condition that reduces the cow's appetite or inhibits her normal behaviour may impact on bulling behaviour. Veterinary intervention to treat the problem(s) is required.

## Heat Detection Aids

A number of devices have been developed to aid in the identification of oestrus. With the possible exception of progesterone sampling kits, they are only aids and should not be seen as substitutes for direct observation.

The most commonly employed technique is to use a product that – when applied to the tail-head of the cow – will register when the cow has been mounted. The cheapest of these are tail paints or markers that are rubbed off during mounting. They have the advantage that they allow the use of colour coding to differentiate between cows that are waiting to be served, those that have been served, and those that are pregnant. The disadvantage of such applications is loss of the paint through natural wear leading to the creation of 'false positive' cows.

The heat mount detector is a device that is glued to the tail-head hair and demonstrates a colour change when rubbed or squashed. Heat mount detectors are generally more robust than tail paints and therefore tend to produce fewer false results, though accidental knocking of the device may occur, for example against a cubicle stanchion.

Pedometers are devices that measure cow movement. A bulling cow shows increased

*Tail paint is a useful way to improve oestrus detection, though there will be false positives.*

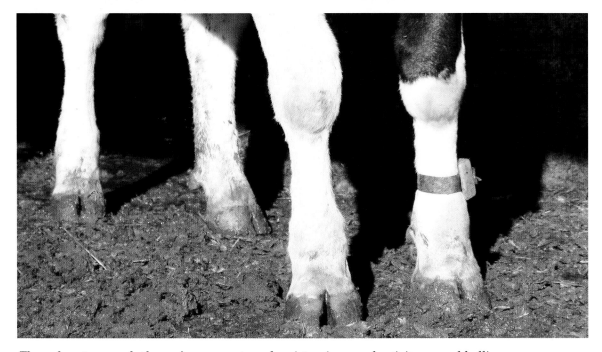

*The pedometer records the cow's movements and registers increased activity around bulling.*

---

**Heat Detection Checklist**

**Keep accurate records**
- Identify which cows need to be observed closely.
- Identify any abnormalities.

**Spend enough time every day looking for bulling cows**
- Ideally twenty minutes three times daily, especially early morning and late evening.

**Make sure cow identity is easily seen**
- Clear freeze brands, collars, eartags.
- Good lighting in housing.

**Increase number of people watching for bulling cows**
- Make sure that everyone dealing with the cows has the skills (and incentives) to spot bulling cows.

**Provide environment for cows to show bulling signs**
- Loafing area.
- Separate group for early lactation cows.
- Avoid:
  Slippery concrete.
  Over-crowding.
  Bullying (first-calved heifers).

**Prevent excess weight loss after calving**
- Cows show signs of bulling more strongly if they are in positive energy balance.
- Tackle:
  Early lactation disease.
  Causes of variation in condition score within the herd.
  Fat cows prior to calving.
  Rapid loss of condition post calving.

**Use aids to oestrus detection**
- Heat detectors.
- Pedometers.
- Teaser bull.

**Use your vet**
- Discuss records.
- Fertility examination to identify problems.
- Pregnancy test to identify cows that are not in calf.

**Minimize stress**
- Ensure smaller groups, better housing, plenty of space for feeding.
- Separate heifer group to avoid bullying.

**Treat lameness early and implement lameness control programme**
- Lame cows lose weight and show no signs of bulling.

---

activity levels, which are registered as activity spikes by computer, flagging the cow for the farmer's attention. They are potentially very useful but require high capital investment compared with more simple devices.

The bull is by far the most adept at correctly identifying a bulling cow, so use of a vasectomized bull (teaser bull) has great potential. The bull can be fitted with a 'raddle', a marker device that sits in front of the bull's chest to mark the cow when he mounts her. A teaser bull run with the cows must be managed appropriately, making sure that he does not gain excess condition owing to access to the dairy ration and ensuring that the environment is suitable to prevent lameness. Slippery surfaces must be avoided because of the risk of injury to bull and cows. Safety hazards that arise whenever a bull is allowed into a working environment must be addressed. An alternative method is to keep the bull in a pen near the cows because bulling cows will tend to congregate close to him.

Progesterone testing can be used to measure daily milk progesterone level, which falls when the cow comes into oestrus. Potentially this is a highly accurate indicator of oestrus, but it may be difficult to interpret results accurately and there is a large degree of hassle in obtaining and testing the samples.

*Artificial Induction and Synchronization of Oestrus*
One potential solution to the difficulties presented when trying to accurately identify the bulling cow is to override nature by using

*Pregnancy diagnosis and veterinary fertility examination help to improve fertility management. (Photo: Kingfisher Veterinary Practice.)*

hormones to artificially induce and/or synchronize oestrus. All require veterinary involvement. Induction of oestrus without synchronization, using a single application of either prostaglandin or progesterone, brings most cows into oestrus within two to six days, allowing a more focused period for observation of oestrus behaviour. Synchronization requires a combination of hormones to be given at specific times and allows cows or heifers to be served at a fixed time after treatment without any requirement for observation of oestrus behaviour.

Synchronization may be essential in suckler herds, or for replacement heifers where artificial insemination is to be used, as observation of oestrus is impractical. The technique allows the development of a tight calving block, but there are increased costs for handling, labour and veterinary treatment. However, provided good pregnancy rates are achieved, these costs are outweighed by the advantages. A variety of different methods to synchronize cows is available, and the options should be discussed with the vet.

## Management of Service and Pregnancy

Following service, it is important to determine whether or not a cow that is not observed to return to oestrus is pregnant. Cows that are ready for service but have not been observed

in oestrus, or demonstrate abnormalities such as vulval discharge or frequent bulling, need to be identified and treated. The use of pregnancy diagnosis in early pregnancy – and the veterinary inspection and treatment of problem cows – is therefore an important component of fertility management.

Many factors influence the success rate of a particular service. They can be aggregated into six main groups:

1. Heat identification.
2. Timing of the service.
3. AI technique.
4. Management of the bull.
5. Nutrition.
6. Disease status.

### Heat Identification

Cows misidentifed as being in oestrus will obviously not conceive to service.

### Timing of the Service

The average time from onset of oestrus to ovulation is approximately twenty-four to thirty hours. It is therefore recommended to serve the cow twelve hours after oestrus is first observed: this will match the timing of service as closely as possible to the point of ovulation and allow time for sperm maturation in the female genital tract. This relies on the assumption that the onset of oestrus is always observed. However, the egg is viable for only four hours after ovulation so it may be better to serve immediately oestrus is observed, particularly where heat detection is sub-optimal.

### AI Technique

AI is carried out either by trained farm staff or by professional operators. Correct preparation and insemination of semen from an AI straw requires care and skill. Poor technique has a serious impact on the conception rate and can potentially harm the cow, so any person performing AI requires training.

If conception rates are sub-optimal, AI technique must be checked. The ability of an individual to successfully insert an AI straw

*Retrieving the AI straw from the flask. The flask contains liquid nitrogen to keep AI straws frozen. Correct storage and handling of straws is essential for reproductive success.*

through the cervix and inseminate a cow is usually best evaluated through the formal checking that is available on a training course. Preparation for use of the AI straw can be checked on farm as follows:

1. Maintenance of the flask.
   AI straws must be kept frozen at a constant temperature. They are stored in an insulated flask, kept cold by liquid nitrogen. The outer vacuum chamber is one of the flask's main insulation components; if damaged, its insulating properties are significantly compromised. Frosting at the top of the flask (caused by rapid nitrogen loss) indicates that damage may have occurred. The flask cannot be maintained at the correct temperature without a sufficient liquid nitrogen, so the level should be monitored routinely and topped up as needed. The flask should be stored in a clean, dry environment (to avoid corrosion) and kept out of direct sunlight and away from other sources of heat.

2. Removal of the straw.
   The straw should be removed from the flask as quickly as possible to avoid unnecessary increase in temperature that might damage remaining straws. Transfer of straws from one flask to another should also be conducted as quickly as possible. A chart should be maintained next to the flask to identify the location of each bull's straws so they can be found easily. The canister holding the required bull's straws should not be lifted higher than the lower half of the neck tube to identify and retrieve the required straw.

3. Preparation of the straw.
   Following removal, the straw should be waved to remove the plug of carbon dioxide at one end and then placed in a warm-water bath for a specified time and temperature, according to the manufacturer's instructions (usually around 37°C for a minimum of forty seconds, but these requirements vary and should be checked for each straw). Following removal the straw should be dried thoroughly. Water is toxic to sperm, so failure to dry the straw may damage the semen.

4. Preparation of the gun prior to insertion.
   Sperm are sensitive to cold, so the gun should be warmed – and kept warm – prior to the insertion of the straw. For this

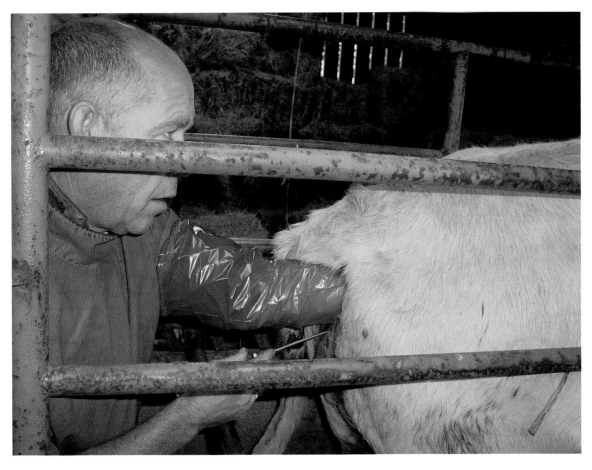

*Inseminating the cow. To enhance conception success rates, and to avoid harm to the cow, this procedure must be undertaken by a properly trained operator.*

reason – and to avoid excessive delay post thawing and prior to insertion – no more than three straws should be prepared at one time. Following insertion into the gun, the tip of the straw should be cut cleanly to provide a straight tip, prior to the insertion of the sheath over the gun and the ring to lock the sheath in place.

5. Handling facilities.

The cow must be suitably restrained, so it is essential that good handling facilities are available. The AI holding pen should also allow the cow to rest, feed and drink while awaiting service, in order to minimize stress.

*Management of the Bull*

This important part of fertility management is discussed later in the chapter.

*Nutrition*

The impact of nutrition on a cow's ability to conceive is complex. Cows in negative energy balance are less fertile than those in a positive balance; the more severe the deficit the greater the impact. Ensuring that cows are in the correct body condition for their stage of lactation and pregnancy is therefore vital in maintaining good fertility. The monitoring of nutrition and condition score is discussed in Chapter 5.

Trace element deficiencies and imbalances may also impact on fertility. The investiga-

---

### Checklist for Poor Pregnancy Rates

**Keep accurate records**
- Identify where things are going wrong.
- Flag up problems early.

**Do not serve too soon after calving**
- Dairy cow may not return to maximum fertility until 56 days after calving (later for suckler cows).
- Negative energy balance is inevitable in early-lactation, high-yielding dairy cows.

**Improve nutrition**
- Conception rate improves if cows are in positive energy balance.
- Tackle causes of variation in condition score within the herd, e.g. fat cows prior to calving and rapid loss of condition post calving.

**Control disease**
- Bovine virus diarrhea (BVD) and infectious bovine rhinotracheitis (IBR) may cause lower conception rates. Vaccinate if needed.
- Bull may carry infections that prevent conception.
- Any post-calving disease is likely to reduce fertility results.

**Use your vet**
- Discuss records.
- Pregnancy test to identify cows that are not in calf.
- Fertility examination to identify problems, e.g. uterine infections.

**Minimize stress**
- Ensure smaller groups, better housing, plenty of space for feeding.
- Separate heifer group to avoid bullying.

**Manage the bull correctly**
- Avoid overuse.
- Treat disease (especially lameness) promptly and do not use a bull that is not fit.
- Test bull fertility – 20 per cent of bulls are not fully fertile.

**Improve timing of AI**
- Make sure oestrus detection is excellent – no cow can conceive unless she is inseminated soon after oestrus.
- Best fertility results are obtained if AI is 6 to 18 hours after signs of standing bulling.
- Make sure AI timing is a priority.

**Make sure the inseminator has adequate skill and time**
- Check records for variation in inseminator performance.
- Refresher course for DIY inseminators.
- Trial period using commercial AI to check conception rates.

**Improve handling facilities for AI**
- Cows need food, water and (ideally) space to lie down while awaiting AI.
- AI is easier if cows are tightly held for insemination.

---

tion of vitamin and trace element status is complex, with many test results difficult to interpret. It requires an assessment of the diet that is fed as well as of the animals themselves, and should be discussed with the herd's vet.

### Disease Status

Disease may directly affect the embryo or reproductive tract to reduce fertility, or it may indirectly reduce the cow's ability to breed, for example by causing a reduction in energy intake. (Lameness is an example of this.) The herd's disease status for both infectious and non-specific diseases is impor-

tant. Intrinsically linked with this topic is the management of the cow at calving (*see* pages 137–141).

## Maintenance and Analysis of Fertility Records

Fertility records are not only important in assisting decisions for managing cows – such as deciding when an individual cow should be dried off prior to calving – but also in understanding how the herd's fertility is performing and where improvements can be made.

Records do not need to be kept on a computer, although, provided software is suitable, analysis and report writing are much easier

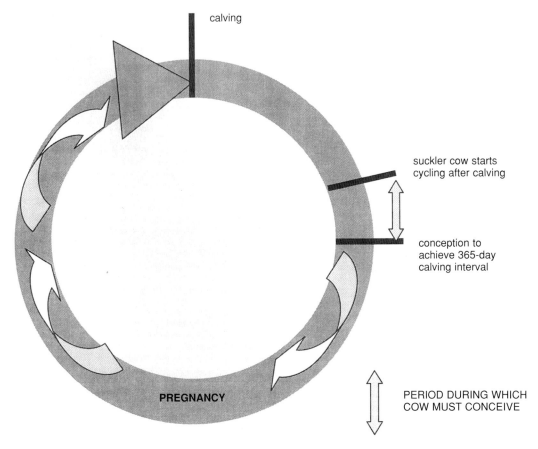

calving

suckler cow starts cycling after calving

conception to achieve 365-day calving interval

PREGNANCY

PERIOD DURING WHICH COW MUST CONCEIVE

*The reproductive cycle of the beef suckler cow. She has only a short period in which to become pregnant if she is to achieve a 365-day calving interval.*

than they are when only manual records are available. Manual records are always the first point of recording, with records subsequently transferred to the computer. The following parameters should be recorded for each individual animal:

- Calving date.
- Dates of observed heat.
- Service dates.
- Person performing insemination.
- Bull used for either AI or natural service.
- Result of pregnancy diagnosis.
- Treatment for reproductive abnormality and the condition treated.
- Date bull(s) enter and leave herd.
- Abortion.

Once collated, the following reports should be generated for the management of the herd and to allow for potential veterinary intervention:

- Cows due for service that have not yet been served – i.e., oestrus not observed.
- Cows due to show a heat – i.e., had a previous heat eighteen to twenty-four days previously.
- Cows due for pregnancy diagnosis.
- Cows due for drying off (dairy cows).

*Beef Fertility Targets*

As discussed in Chapter 2, the prime target for any suckler cow farmer is to maximize the kilograms of beef sold per year at the minimum cost. This requires not only that as

## Table 21: Target condition score for suckler cows

|  | Mating | Mid-gestation | Calving |
|---|---|---|---|
| Spring calving | 2.5 | 3 | 2.5 |
| Autumn calving | 2.5 | 2 | 3 |

many cows as possible calve each year, but that they calve within as short a time frame as possible. This allows feed rates to be tightly controlled, thereby reducing the costs of maintaining the adult herd, and also ensures that all calves have as much time as possible to gain weight prior to being weaned or sold.

The average length of pregnancy is around 283 to 285 days, which leaves only ninety days post calving for the cow to prepare herself for service and become pregnant again. The suckler cow has very little time to become pregnant if she is to maintain a calving interval of 365 days.

The factors determining success will be correct management of:

• Nutrition, specifically ensuring the cow is at the correct condition score at every stage of the reproductive cycle.
• The bull.
• Heifer replacements.
• Disease/calving problems.

A target can be set for each of these parameters: for example, condition score or weight gain in heifers. The targets set out in Tables 21 and 22 demonstrate the required overall performance of the herd to achieve a compact nine- to ten-week calving period with a high percentage of the herd pregnant.

### Dairy Fertility Targets

Within any system of dairy production, the maximum efficiency of milk production is gained if the highest possible percentage of the herd calves annually, with a minimum number culled for infertility.

Fertility records tend to be more detailed in dairy than in beef herds because dairy cows

### Table 22: Fertility targets for suckler cows

| Parameter | Target |
|---|---|
| Duration of calving period | 9 weeks |
| Calves weaned per cow served | >94% |
| Percentage of cows pregnant within the first three weeks of service | 65% |
| Percentage of cows pregnant within the second three weeks of service | 20% |
| Percentage of cows pregnant within the third three weeks of service | 10% |
| Adult cows with dytocia | <5% |
| Heifers with dytocia | <15% |
| Calf loss (abortion/stillbirth and death soon after birth) | <4% |
| Replacement rate | 15% |

can be more closely observed. This provides an overview of fertility performance and allows more complex analysis to investigate problems. As with the beef herd, an analysis should be made of the other factors that influence fertility, such as nutrition and disease.

## MANAGEMENT OF THE BULL

Bulls can be used alone, as is the case in the majority of suckler units, or in combination with AI as a means of 'sweeping up' animals that do not fall pregnant to AI. The use of a

---

## Table 23: Fertility targets for dairy herd

|  | *Target* |
|---|---|
| **Analysis of heat detection performance** | |
| Calving to first service interval | 65 days |
| First service 24-day submission rate* | >75% |
| Inter-service interval analysis** | *See* below |
| | |
| **Analysis of pregnancy rate*** | >50% |
| | |
| **Analysis of overall herd performance** | |
| Calving interval | 365 days |
| Number of cows served that become pregnant | >92% |
| Number of cows culled for infertility | <10% |
| Percentage of cows in calf by 100 days | >75% |
| Percentage of cows not in calf by 200 days | <10% |

\* This measures the number of cows served within 24 days of the start of their service period. Thus if cows are to be served from 60 days post calving it will measure the number of cows that actually were served up to 84 days from their calving date.

\*\* This index calculates the number of days that lapse between each service a cow experiences and then groups them into defined intervals. The targets are shown in the graph below.

\*\*\* There should be no variation in pregnancy rate linked with service number, whether bull or AI, days post calving, inseminator, yield, or day of the week.

---

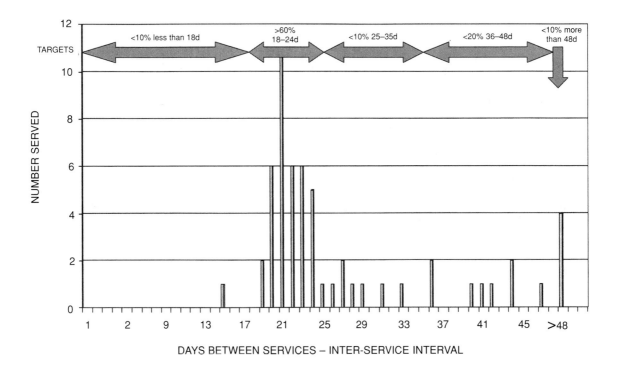

*Inter-service intervals.*

bull/bulls to serve cows rather than the use of artificial insemination provides many advantages, but also many disadvantages. These are outlined in Chapter 4.

The correct management of bulls is crucial, as one sub-fertile/infertile bull will lead to a potential failure to get many cows in calf, with severe financial consequences. The important considerations to be taken into account are explored below.

## Genetics

In choosing a bull, one should bear in mind the likely ease of calving, and the bull's ability to produce calves suitable for the chosen market. An inappropriate bull can cause calving difficulties and/or the production of inferior calves. These are inherited characteristics that are discussed later in the chapter. Until a young bull starts producing his first crop of calves, it may be difficult to judge whether or not he is suitable because the only information available (if any) is derived from his parents.

## Nutrition

The bull must be in the correct condition at the start of a service period if he is to perform optimally.

The young bull must achieve an adequate body size and condition. Bulls from the age of six to eight months to the age they are first used – at eighteen to twenty months – need to gain about 1kg daily for British breeds, or 1.2kg daily for larger Continental breeds. This ensures that they reach a weight of about 600kg. Weight should be monitored monthly to assess progress.

From the first point of use until maturity at thirty to thirty-six months of age (dependent on breed), weight gain should be reduced to 0.8–0.9kg per day. Care should be taken to ensure that lean weight is gained rather than excessive fat. If bull condition increases, the concentrate component of the diet should be reduced and the protein level of the diet checked. Diets should contain about 12 per cent crude protein. Good-quality grazing is often sufficient nutrition for this age of bull.

From maturity, feeding should be based on the maintenance of correct condition, particularly up to and at the point of breeding. A target condition score of 3 to 3.5 should be attained at the onset of breeding. During the breeding season bulls will lose condition.

## Housing

Outside the breeding season, bulls in beef herds must be kept separate from cows. They should have enough space to exercise. If bulls are managed in groups they should be of similar ages and have been reared together: this will aid feeding management and reduce fighting. Grazing is the ideal system as it provides for both the nutritional and physical requirements of the bull. A bull is a very strong animal, so fencing must be robust enough to prevent escape (usually in pursuit of bulling females).

Where bulls are run continuously with a housed dairy herd, provision should be made to rotate and rest the bulls. This prevents the development of foot problems owing to inappropriate housing conditions, and overfatness owing to ad lib access to the dairy diet. Ideally, resting bulls should be housed in a straw yard. They should be in view of the herd and working bull(s) in order to increase the sense of competition between them – thereby raising their libido – and to allow younger bulls to observe and learn.

In herds where bulls work together, there is always the risk of fighting. Ideally, if more than one bull is required, each should have a separate paddock. If this is not feasible, bulls of differing ages should not be run together as the young bulls may be bullied and prevented from working by the more dominant older animals.

## The Health Status of the Bull

Bulls are frequently implicated in the importation of a variety of diseases into herds. This is particularly true of hire bulls but can apply to any bull. The importation of a sexually transmitted disease, such as campylobacter,

*Beef bull running with his herd. Note that he has a nose ring as an aid to management.*

can have a devastating impact on a herd's fertility. Virgin bulls should be purchased from herds of known disease status wherever possible. Where disease status is not known, tests should be performed prior to the purchase.

On arrival, the bull should be isolated for six weeks. During this time appropriate tests will establish the individual's disease status and allow interventions or treatments both to protect the herd from diseases the bull may carry, and to protect the bull from diseases in the herd to which he has no immunity.

Foot problems, arthritis and back pain inhibit the serving capacity of the bull. Regular assessment of his gait and service should be made and his feet regularly trimmed.

## Inbreeding and Unwanted Pregnancies

It is essential to make sure that the bull does not breed with his own daughters: this can be achieved either by the purchase of heifer replacements, or by replacing the bull every two years. In pedigree units it is usually necessary to create separate groups for different lines to avoid inbreeding.

If heifer and bull calves are kept together, or the adult bull runs with cows and calves, there is a high risk of unwanted pregnancies. Bulls can be fertile from as early as six months, (though normally fertility will occur at around twelve to fifteen months), and heifers may become pregnant from four to six months old. Care should be taken to avoid this.

## Fertility Status and Serving Capacity

Approximately 20 per cent of bulls are sub-fertile or infertile, i.e., their semen quality is impaired. Sub-fertility may be further exacerbated by abnormalities of the penis, painful musculoskeletal problems, or even psychological problems that reduce the ability or desire to serve a cow. A thorough examination and assessment of the bull's fertility and capacity to serve should be made before he is bought or leased. This should be done even when a previous service history is available, as circumstances can change and the economic impact of an infertile bull can be huge. This breeding soundness assessment should be made prior to each breeding season.

A basic assessment of the bull's potential fertility status includes examination of the following:

- Conformation and soundness. Identify abnormal gait or lameness, sickle-shaped or excessively straight hocks, poor claw conformation.
- Testicular structure. Examine size, structure and uniformity of the testicles. A measurement of the scrotal circumference, at the maximum diameter, is a very useful guide to fertility status, particularly in young bulls. A

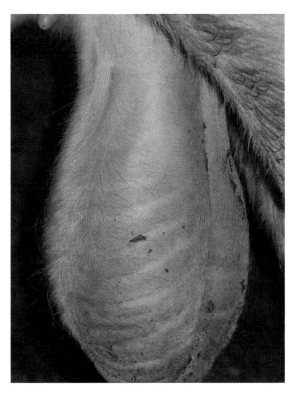

*The testicles should be large, firm, even-sized, and freely mobile within the scrotum.*

### Table 24: Minimum scrotal circumferences

| Age (months) | Minimum scrotal circumference |
| --- | --- |
| >18 | 32cm |
| >24 | 34cm |

circumference of 32cm is the minimum requirement. The testicles should hang freely from the body, move freely within the scrotum, be of similar size, and should feel moderately firm on palpation (similar to that of an avocado at the point of ripeness).

The most accurate assessment of bull fertility can be made only by a combination of semen evaluation and observation of service. The evaluation of semen has become much easier, safer and more economical since the

*The capacity of the bull to serve should, if possible, be observed.*

advent of electro-ejaculation technique. This is performed by a veterinarian and relies on electric charge discharged from a rectal probe to artificially cause ejaculation. It does not require a bulling female and allows a much safer collection of semen. The disadvantages are that the volume of semen collected tends not to be as great as it is with natural ejaculation and it does not allow observation of the bull's ability to serve.

The capacity of a bull to serve the required number of cows is a function of:

• The age of the bull. The ratio of cows to a single bull based on age is outlined in Table 25. A good average is one adult bull to twenty-five to thirty cows.

• The workload of the bull. A bull that has been working continuously will have a far lower capacity to serve than a bull that is rested and fresh. A fully mature bull running for a short period (four to six weeks) is generally capable of serving 2–2.5 cows per day; over longer periods the same bull is capable of serving 0.5–1.0 cow per day.

• The environment in which the bull is maintained. A bull running with a group of cows in a field is far less prone to wear and tear than a bull running with cows in a cubicle house with access to a lactating cow ration.

| Table 25: Ratio of cows to a single bull | |
| --- | --- |
| Age of bull (months) | Number of cows per bull |
| 15–18 | 12–15 |
| 24–36 | 25–30 |
| >36 | 40* |

\* Provided the breeding season is short

*A Hereford bull. A bull that has been well handled is less of a risk when out with the cows.*

- The ability to rotate the bulls. The frequent rotation of bulls between work and rest has the advantages of preventing excess loss of condition, avoiding a reduction in serving capacity and libido (the desire to serve), and decreasing the chances of injury or lameness. Whether within the beef or dairy herd, bulls should, ideally, be rotated every two to three weeks, though there is flexibility according to the number of cows for service and the serving capacity of the individual bulls. Some bulls are inherently capable of maintaining their serving capacity for much longer than others. Bulls that are performing well can be kept longer with the herd.
- Whether the bull will be serving synchronized cows/heifers. If the bull is being used to serve animals that have failed to hold to a synchronized AI service, it is likely that returns to service will occur within a short time frame. The number of bulls used will need to reflect this.

## Observation and Records

For many farmers, the advantage of a bull is that it releases time and labour from the management of the service. This should not be seen as an excuse for a total abandonment of fertility management, however; observation of the bulling behaviour of the cows to ensure that not too many cows return to heat and that the bull is successfully mounting cows should still be maintained. The use of a raddle on the bull to mark cows that he serves can be a useful aid, both to ensure cows are being served and to provide a reference, based on the colour of the raddle, for a return to service and the likely date of calving. Unfortunately some bulls are inhibited by the raddle's use.

## Safety Considerations

An adult bull is a potentially dangerous animal, and he should be treated with respect and caution. The bull should be kept in a manner that will reduce the potential danger to the farm personnel, to members of the public if at pasture and, in certain instances, to the cows themselves. There are legal requirements, pertaining to the keeping of a bull, which should be observed (*see* Chapter 10).

## THE CALVING COW

The length of a cow's pregnancy ranges from 280 days to 290 or more, depending on the breed and other factors. While the timing of delivery is determined by the calf via release of the 'stress' hormone cortisol, several indicators of approaching parturition (calving) can be observed in the cow, such as increased mammary development, gradual production of a milk precursor that becomes colostrum, and relaxation of the pelvic ligaments.

The calving cow requires a clean, dry comfortable environment free from disturbance (*see* Chapter 8). Plentiful clean water and palatable feed should be provided close to the cow so she can eat and drink without moving far from her calf.

The process of calving is divided into three stages.

## Stage 1: Onset of Parturition

At the onset of parturition, behavioural changes in the cow can be observed. These may include restlessness, seeking solitude, not eating or eating little, rising and lying down, looking round at the belly, and twitching the tail or holding the tail high. During this stage, labour progresses in the following way:

- The relaxation of the pelvic ligaments reaches maximum.
- Contractions of the uterus begin and become more frequent.
- The cervix begins to open. The degree of cervical dilation (ascertained by vaginal examination) is a good indicator of the progress of Stage 1. The thick, sticky, mucous plug that blocks the cervix is released.
- The cow's temperature may fall by one or two degrees centigrade; respiration and heart rate may rise.
- Continued contractions of the uterus move the calf towards the cervix.
- The water bags appear (allantois and amnion that surround the calf), further dilating the cervix.

Stage 1 lasts from thirty minutes to twenty-four hours, but it is difficult to determine its onset as uterine contractions are not visible. Any stress can delay or even stop the process, so cows should be left alone. If they must be handled this should be done quietly and sympathetically.

## Stage 2: Delivery of the Calf

During the latter stages of Stage 1 the calf is pushed up and into the pelvis. This is an active process during which the calf aligns itself into the correct position for birth. This is the onset of Stage 2, which will proceed in the following way:

- The presence of the calf in the pelvic canal stimulates abdominal contractions. These gradually become more forceful and the cow can be observed grunting or blaring as the contractions develop. The cow is now 'straining' with contractions that are occurring four to eight times in ten minutes.
- The water bag breaks.
- The calf appears at the vulva. It may be presented either with front legs and head first (anterior presentation) or with hind legs first (posterior presentation). Both are normal. Anterior presentation is far more common, and passage of the head through the pelvis is the time of maximum effort for the cow. The cow often rests after the head has been passed and then resumes straining until the thorax is delivered. The hips and hind legs are usually passed with much less effort. With a posterior presentation, the passage of the hips is the maximum effort for the cow. Posterior presentation is often seen in the delivery of twins.

Stage 2 lasts one to four hours, with heifers taking longer than cows. Provided progress is being made, interference is unnecessary. If there is no progress after two hours of Stage 2, there may be a calving problem (dystocia) and the cow should be examined.

## Stage 3: Delivery of the Placenta

This is the final stage of labour, which begins after the calf has been born. During this stage, the foetal membranes (the placenta – also called the afterbirth or 'cleansing') are pushed out. For the delivery of the placenta, the cow will usually have stopped straining, but the involuntary uterine contractions continue in order to push out the membranes.

This stage should last no longer than six hours. Continued straining after this period may indicate that there is another calf so the animal should be checked.

*First-stage parturition:*

*1. Heifer in first-stage labour looks round at her belly in pain.*

*2. The tail is cocked out and a string of slime appears from the vulva.*

*3. Water bag appears at the vulva, indicating that the cervix is opening.*

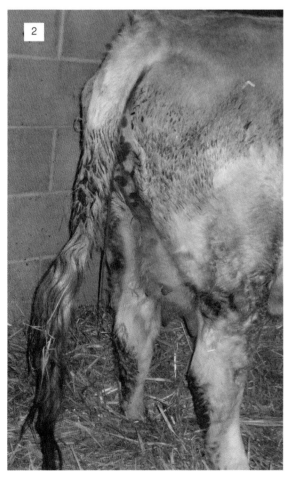

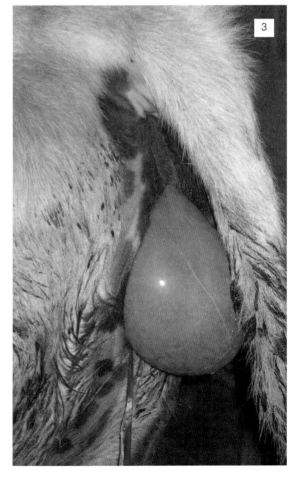

***Second-stage parturition:***

*Normal calving with calf in anterior presentation. The calf is still covered with the membranes that aid its passage through the birth canal. Forefeet are extended in a diving position, with the nose about midway between fetlocks and knees.*

## Dystocia

Difficulty in delivering the calf – termed dystocia – can affect both the beef and the dairy cow. Dystocia is most likely to be caused by one of the following:

- A relative disproportion between the calf's size and the mother's size (specifically her pelvic size).
- An abnormal presentation or abnormality of the calf itself.

Relative disproportion between the calf's size and the mother's size is by far the most common problem. The classic causes are:

- Inappropriate choice of bull for the breed or age of female. When selecting a breeding bull, the likely size of calf compared to the breed and age of dam must be considered. Where available, genetic selection indices for individual bulls for the average gestation length (the shorter the better) and ease of calving should be used. These are available for beef bulls in the form of estimated breeding values (EBVs), discussed on page 143.
- Inappropriate management of the heifers, either by serving them when they are too small or by failing to achieve target liveweight gain post service. (Correct management of the heifer is discussed in Chapter 4.)

An abnormal presentation or abnormality of the calf itself can be seen as a random event, or it may result from the calving of twins or an oversized calf, or from conditions such as milk fever. The correction of an abnormal presentation requires considerable skill and understanding of cow and calf anatomy. It should be performed only by trained personnel.

*Intervention*

The indication that there may be a problem is the failure to progress after two hours of Stage 2 labour. If a problem is confirmed on examination it may still be possible to deliver a live calf with the aid of lubrication and sensible traction.

Determination of the ability and readiness of the cow to pass her calf can be made by a vaginal examination to make sure that the cervix is fully open, and, where necessary, by attempting reasonable traction on the calf using calving ropes. Making this assessment requires experience and training.

Before intervening, the following questions should be considered:

1. Is the cow due to calve?

2. Has she started straining?

3. If so, for how long? When was she last examined and by whom?

4. Is she showing clinical signs that need treatment (for example, milk fever)?

A full assessment of the calving cow is impossible without adequate restraint to ensure a safe environment for both the operator and the cow. A dedicated calving box with handling facilities is ideal. A crush should only ever be used to assess the cow, never to calve her. All required equipment should be to hand, for example two buckets of clean warm water, disinfectant, plenty of lubricant, ropes (or chains) for feet and head, two short sticks to pull the ropes, or a calving aid.

The cow's back end should be thoroughly washed and the operator's hands and arms should be thoroughly cleaned using disinfectant. Plenty of lubricant should be used and a hand inserted into the vagina to make an initial assessment in order to answer the following questions:

1. Is the cervix fully dilated?

2. Are any feet present? If so, are there two forefeet and a nose, or two hind feet and a tail?

3. Do all the parts belong to the same calf?

4. Is there torsion of the vagina?

Again, skill and experience are required to identify whether or not the calving is proceeding normally. If the cervix is not yet dilated and the calf is not yet presented in the pelvic canal, the cow needs more time and should be left quiet and monitored. If the calf is normally presented in either anterior or posterior presentation, she can be allowed to continue to calve. However, if the calf seems excessively large, if the cow becomes distressed or exhausted, or if an abnormal presentation is felt, intervention will be needed.

If the calf must be delivered by traction, a calving aid (a machine that applies traction) should generally not be used as it is easy, in inexperienced hands, to apply too much traction and cause injury to cow or calf. Calving aids should be seen as devices for saving energy or labour; they are not intended to winch a calf out of a cow with more force than could be provided by two normal adults. Ultimately the skill to calve cows is learnt from experience and practical teaching. It is always better to call for assistance than risk the loss of calf, cow or both.

Following any assisted birth, the cow's birth canal should be checked for damage such as bleeding or vaginal tears. The cow should also be checked for the presence of another calf.

## Aftercare

It should be remembered that cows can be very calf-proud and threaten an intruder who approaches, so care must be taken if cow or calf must be examined after calving.

If the cow continues to look uncomfortable or to strain, she should be examined for damage or the presence of a second calf. If

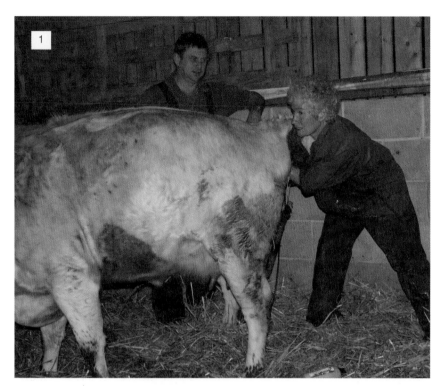

**Second stage parturition:**

1. Having made no progress for four hours, this heifer is examined. The calf's head is back and delivery cannot proceed until it has been brought into the right position.

2. After correction of the malpresentation, two people pull the calf. A live heifer calf is delivered with no damage to the mother.

(Photos: Ulrike Wood.)

the cow is recumbent, she should initially be allowed to rest. However if she remains recumbent there may be damage to nerves or muscles, or she may be suffering from milk fever. Veterinary attention should be sought.

# PRINCIPLES OF GENETIC SELECTION

Genetic selection is a powerful tool for ensuring the presence of specific characteristics in the animals that are bred. If genetic selection is used to make changes in the herd in order to increase herd profitability, selection must take account of:

- The likely demands of the market.
- Farm resources such as feed quality, housing, and skill and availability of labour. Selection for increased production increases profits only if the animals remain healthy and productive. For example, there is no point in breeding animals capable of achieving very high milk yields if the farm uses a low-input grazing system, nor in using a large, double-muscled bull with cows that must calve on the hill without supervision.

In the beef sector the bull is chosen to produce calves that will mature to finished animals that meet the demands of the market, combined with selection for traits such as ease of calving and feed conversion efficiency. In the dairy sector the choice is based on the desired level of yield and constituent quality, and health traits such as longevity, udder and foot conformation.

Most genetic gain is achieved by using AI bulls. However, it is also important to select cows that show the desired traits and to choose an AI bull that counteracts any weakness of a particular cow (e.g. a bull for improving udder conformation matched with a cow with poor udder conformation). Embryo transfer is an advanced reproductive technique where an exceptional cow is treated with hormones so that she produces multiple ova that can be fertilized, and the embryos grown in other cows.

## Dairy

A calculation is made of an animal's genetic index – a measure of its ability to transmit its genes to the next generation. It is expressed as a predicted transmitting ability (PTA). A bull's PTA index is calculated from the performance of his daughters (progeny tests) and, to a lesser extent, other relatives. For a cow the index includes her own performance. The data collected is adjusted to take account of management variations in the herds carrying the progeny of the animal. The animal is then compared to a national fixed-base standard. For instance in the UK the current base is termed PTA2000 and, for a heifer, equivalent production is set at 6,279kg milk, 251kg fat, 206kg protein, 3.98 per cent fat, and 3.26 per cent protein.

The more information that is available from the offspring of an individual the more accurate the PTA will be. This potential accuracy of the data is called the 'reliability of the proof' and is expressed as a percentage, with a higher percentage indicating more reliable data. The minimum for bulls is 30 per cent; ideally, however, only bulls whose reliability is 90 per cent and above should be selected. If bulls with low reliabilities are used (semen from such bulls is cheaper), it is always best to use several different bulls to spread the risk of 'failure'.

PTAs are primarily aimed at demonstrating the amount of liquid milk and milk constituents that an animal is likely to transmit to its progeny. However, high milk production is not the only factor to influence profitability, and PTAs now exist for factors associated with longevity and health.

The calculation of PTAs also allows a calculation of the likely economic advantage of using a particular bull or cow. In the UK, the profitability index number (PIN) is an economic index calculated from milk, fat and protein PTAs and is the expected difference in net margin per cow per lactation.

*Genetic selection for conformation produced these even-sized young bulls.*

## Beef

Beef genetic indices use very similar methods to the dairy models. The animal's genetic index is calculated as the estimated breeding value (EBV) from two sets of factors:

1. Beef value (conformation and growth rate).
2. Calving value (ease of calving).

Data is collected from all recorded relatives and includes information such as:

- Gestation length.
- Ease of calving.
- Birth weight (which is positively correlated with weights at 200 and 400 days, but negatively correlated with ease of calving).

- 300-day weight.
- 400-day weight.
- Fat depth.
- Muscle depth.
- Muscle score.
- 200-day milk weight.

Each of the EBVs is weighted for economic importance and combined to give a numerical index for calving value and breeding value. The traits that are measured vary between countries and organizations. Selection for one trait may adversely affect another; for instance, as weaning weight is correlated to birth weight, selection for increased weaning weight may result in increased dystocia due to large calves.

# CHAPTER 8

# Housing

A range of housing types has been developed, from ramshackle sheds built from recycled materials to state-of-the-art constructions costing hundreds of thousands of pounds. But the quality of a building depends more on how well it serves its purpose than on how much it costs, and this can be gauged in terms of animal comfort and performance, ease of use, and level of housing-related disease. This chapter examines first the principles of cattle housing and then some specific factors for meeting the needs of different classes of stock.

## PRINCIPLES

The main purposes of housing cattle are to shelter the animals from bad weather and thus maintain welfare and production, to keep them off the land (protecting it from poaching), and to enable the stockperson to feed, manage and observe stock easily. The most important factors that determine whether or not these purposes are achieved are discussed below.

### Space Allowance

Clearly cattle need enough space to stand up, lie down, eat and drink; and – because they are herd animals – they should be able to perform these activities at more or less the same time. For example, if the dairy herd is observed an hour after milking, almost all will be lying down, resting and chewing the cud; a group of beef cattle will mostly choose to feed at the same time, even if food is constantly available.

Group-housed cattle also require enough space to avoid injury or stress: the first calved heifer needs space to avoid large dominant cows; the calving cow needs to retreat to her own space to calve and then to mother her newborn calf.

Table 26 (*see* page 146) gives specifications of minimum space requirements for cattle of different classes and weights. These are minimum requirements: in general, the more space available the better.

Problems that are commonly associated with excess stocking density or inadequate trough space include:

1. Uneven growth rates. Competition for trough space prevents smaller calves from accessing sufficient feed, so food intake is reduced and they fail to thrive. (There are, of course, many other causes of ill-thrift in calves, but this simple problem is remarkably common.)

2. Build-up of infectious organisms leading to diseases such as diarrhoea in calves and environmental mastitis in dairy cows. Incidence of these diseases is strongly linked with the number of infectious organisms in the bedding, so crowded housing is a common causative factor. Diseases are also linked with drainage and management of bedding, so high disease incidence is not an inevitable consequence of high stocking density, provided that bedding is kept clean and dry. Conversely, environmental

*A ramshackle shed, though not ideal, provides basic shelter for a small number of calves and keeps them separate from other stock.*

## Table 26: Space allowances for cattle of different classes

| Class of stock | Minimum space allowance per head bedded pens (m²) | Minimum loafing space allowance per head' (m²) | Slats (m²) | Feed trough per head (mm) (figures in brackets for stock fed ad lib) |
|---|---|---|---|---|
| **Calf:** | | | | |
| 4 weeks | 1.1 | – | – | -350 |
| 12 weeks | 1.8 | – | – | 350 |
| **Youngstock:** | | | | |
| 200kg | 3.0 | 2.5 | 1.1 | 425 (220) |
| 300kg | 3.4 | 2.5 | 1.5 | 525 (270) |
| 400kg | 3.8 | 2.5 | 1.8 | 600 (300) |
| 500kg | 4.2 | 2.5 | 2.1 | 650 (325) |
| 600kg | 4.6 | 2.5 | 2.3 | 675 (350) |
| 700kg | 5.0 | 2.5 | 2.5 | 700 (350) |
| **Cow** | | | | |
| <650kg | 5.5 | 2.5 | 3.4 | 700 |
| 650kg and over | 6.0 | 3.0 | 3.6 | 750 |
| **Bull pen** | 16.7 | 32 | | |

*State-of-the-art housing for the dairy herd represents a large investment.*

diseases may be encountered even at low stocking rates if hygiene standards are inadequate.

3. Bullying. First-calved heifers in dairy herds are particularly vulnerable to bullying (*see* Chapter 5). If straw yards or cubicle sheds are overcrowded, the bullied heifer spends less time lying down, and may be forced to twist and turn to avoid confrontations with dominant cows. These factors can result in lameness and solar ulceration.

4. Mismothering. The suckler cow that calves in a crowded loose yard may have her calf 'stolen' by other heavily pregnant animals. In some cases this may mean that the calf fails to suck colostrum.

5. Buildings cannot be 'rested' in between periods of use.

6. Isolation facilities are not available to house sick animals away from healthy stock.

## Temperature

With the exception of young calves and sick animals, cattle are able to regulate body temperature within a relatively wide range of ambient temperature (the so-called thermo-neutral zone, between the lower and upper critical temperatures), and, provided cattle are healthy and have a dry and draught-free bed, they do not need to be kept warm. There are health, welfare and production costs if temperature is outside the thermo-neutral zone: if it is too cold, cattle must burn energy simply to maintain core body temperature; the body temperature of a sick calf may drop so that it is unwilling to feed and drink; heat stress may decrease appetite and, if severe, cause heart failure. Heat stress is much more likely to occur where relative humidity (RH) –

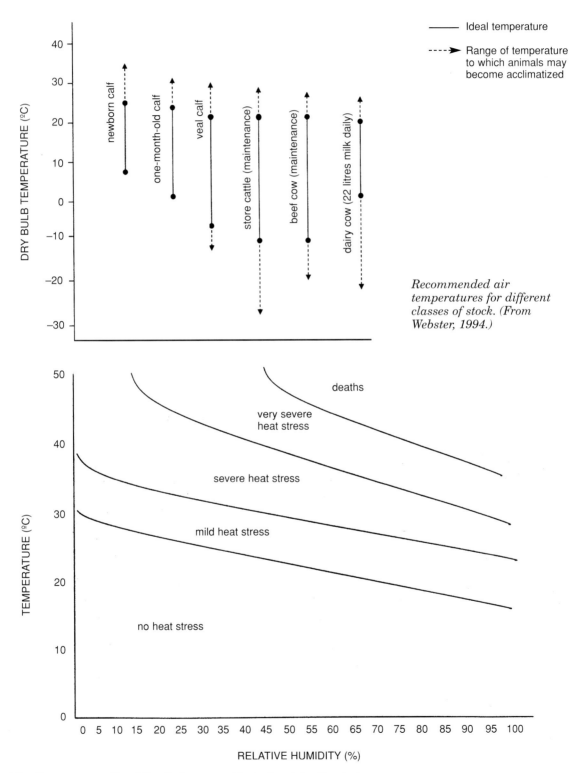

*Recommended air temperatures for different classes of stock. (From Webster, 1994.)*

*The Temperature–Humidity Index – a guide to the risk of heat stress.*

*Outwintered Highland heifer whose thick coat provides insulation against cold.*

*see* below – is high, and growing beef cattle and dairy cows are particularly susceptible. Heat stress reduces milk yields and may cause a decrease in fertility in dairy cattle.

Air inlets into a building should never be reduced in an attempt to maintain temperature raised in winter because this is likely to allow build-up of respiratory pathogens, moisture, and toxic gases, and increase the risk of pneumonia.

## Relative Humidity

The moisture content of the air is measured as relative humidity (RH). Harmful bacteria and viruses (for example those causing pneumonia) survive well in air with a high RH, so the aim should be to keep RH as low as possible. This can be done by increasing ventila-

tion and improving drainage to remove urine and dung, and by ensuring that cattle are not heat-stressed and sweating. The use of hose-pipes in buildings should be kept to a minimum, water trough leaks should be repaired, and rain should be excluded. The use of breathing or slotted roofs (*see* below) can reduce the risk of condensation.

The link between temperature and RH is quantified in the Temperature–Humidity Index, which identifies environmental conditions that put animals at risk of heat stress. These are classified in various ways, perhaps most usefully as: no stress; mild stress; severe stress; very severe stress; and death – *see* diagram opposite (bottom). The picture is further complicated by sunshine, which increases the risk of heat stress, and wind speed, which reduces the risk. In order to minimize the risk of heat stress, cattle should have access to shade, plenty of fresh cool water, a large air space (high roof), and good ventilation. Cattle should not be transported or moved in the heat of the day.

## Ventilation

Ventilation in cattle housing not only supplies fresh oxygen for cattle to breathe, but it removes toxic gases such as ammonia (from urine), excess moisture (from dung, urine and respiration), and pathogenic viruses and bacteria shed from the respiratory tract. Ventilation is determined by the air space available and the rate of air flow, and in a well-ventilated building the air is fresh, dry and draught-free.

Cattle ventilation requirements are determined by the weight of the animal and also by the animal's metabolic rate. For example, a high-producing dairy cow or rapidly growing bull beef animal is metabolizing quickly with relatively rapid respiration and large output of moisture (up to 22 litres per day from skin, up to 9 litres per day from the respiratory tract, and up to 50 litres urine and faeces per day). Such an animal requires more ventilation than a barren beef cow of similar bodyweight at the end of her lactation. The ventilation requirements of various classes of stock are set out in Table 27 (*see*

below). Maximum air speed at stock level should be 0.25m per second.

It is relatively straightforward to calculate the air space available within a building. It is much more complicated to estimate the rate of air flow or ventilation rate. In the winter, stock buildings need a minimum of six air changes per hour to keep the air fresh. In the summer this increases to at least twenty air changes per hour to help prevent heat stress. Ventilation depends upon the fact that livestock produce heat and that warm air rises (the so-called stack effect). Moist air also rises because water vapour is less dense than air, so as air picks up moisture from the cattle it rises. Ideally, air enters the building at a relatively low level, circulates around the stock, gradually warming and rising, and exits from an outlet at the ridge. In a conventional building with a pitched roof, air enters at the sides and leaves through the roof, while in an open-fronted building air enters and leaves by the same route.

Air flow is therefore determined by a number of factors including:

1. Dimension and position of air inlet(s). The inlet area should be at least twice – and preferably up to four times – the outlet area, with minimum air inlet being 0.1m² per calf, 0.2m² per cow. This may simply be the open side of a house, and is

---

### Table 27: Optimum environmental conditions for cattle

| *Stock* | *Live weight (kg)* | *Minimum air space allowance per head (cu. metres)* | *Ventilation rate per kg liveweight\* (cu. metres per hour)* |
|---|---|---|---|
| Calves 0–12 weeks | 50–200 | 5–12\*\* | 0.75–2.25 |
| Youngstock | 200–400 | 12–16\*\* | 0.25–1.25 |
| Fattening beef | 400–700 | 18\*\* | 0.2–1.5 |
| Adult dairy cows | 400–800 | 35 (cubicles)\*\*\* 40 (straw yards)\*\*\* | 0.2 – 1.5 |

\* From Buildings and Structures for Agriculture BS5502
\*\* From Hartung, 1994
\*\*\* From MDC 2006 (a)

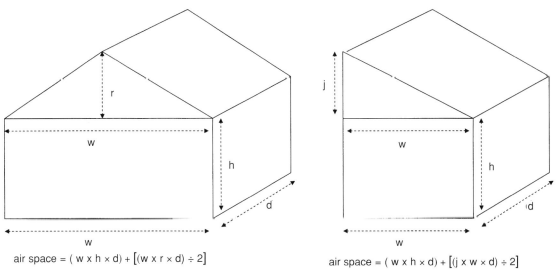

air space = ( w x h × d) + [(w x r × d) ÷ 2]

*Calculation of air space for a double-pitched building.*

air space = ( w x h × d) + [(j x w × d) ÷ 2]

*Calculation of air space for a monopitch building.*

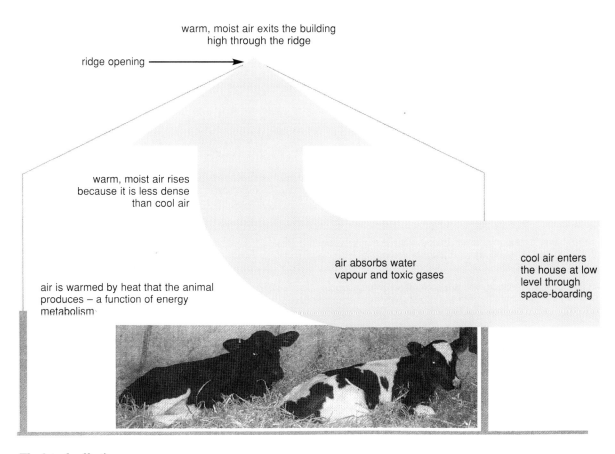

warm, moist air exits the building high through the ridge

ridge opening

warm, moist air rises because it is less dense than cool air

air absorbs water vapour and toxic gases

cool air enters the house at low level through space-boarding

air is warmed by heat that the animal produces – a function of energy metabolism

*The 'stack effect'.*

*Space-boarding baffled with net against the prevailing wind.*

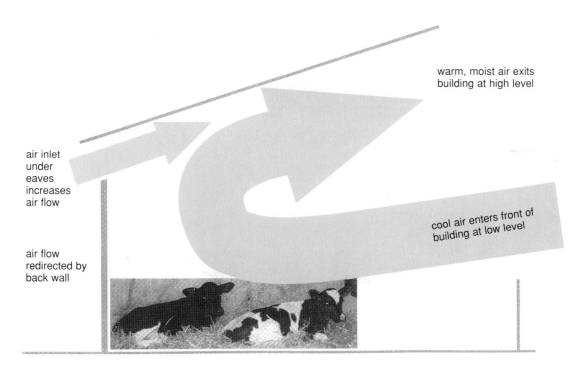

warm, moist air exits
building at high level

air inlet
under
eaves
increases
air flow

cool air enters front of
building at low level

air flow
redirected by
back wall

*Schematic diagram of air flow in a monopitch building.*

often provided by space-boarding, which excludes most wind and rain. There is generally a solid wall with space-boarding above. Ventilation across a single-span building will be more effective if an additional air inlet is created at the back.

2. Dimensions and position of air outlet(s).
The outlet area should be half the inlet area (minimum 0.05m² per calf, 0.1m² per cow), and this may be the side of the building. Alternatively, ridge ventilation – an opening at the highest point of the roof – may be used. An open ridge is cheap and effective, but does not exclude all rain, (width is generally 3–4cm for adult cattle or 5cm ridge per 3m building width). Capped ridges, for larger buildings, have a small roof over the ridge, which keeps out rain or snow. The capping must be de-

signed so that it does not affect air flow out of the ridge. Slotted roofs are used for wide-span and multi-span buildings or where it is difficult to create air outlets. The slots should be a maximum of 12mm wide. This provides an effective air outlet, without rain entry. It is the heat from the livestock that prevents rain from entering, so buildings with slotted roofs are unsuitable for storing hay or straw.

3. Width of the house.
The principles of natural ventilation are not valid in very wide and multi-span buildings, (many modern buildings have a span of 60m or more), so ventilation may be rather hit and miss, with 'dead spots' of poor air circulation in the middle of the building. Slotted and breathing roofs prevent dead spots.

*Open-ridge ventilation.*

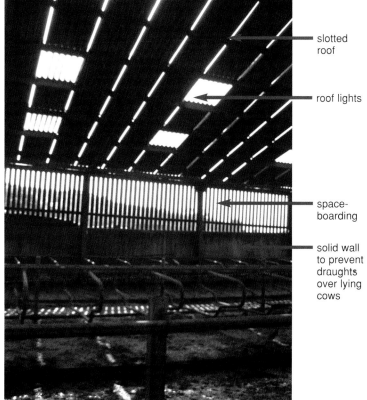

slotted roof

roof lights

space-boarding

solid wall to prevent draughts over lying cows

*Good ventilation and natural lighting in cubicle house.*

4. Pitch of the roof.
   This should be at least 12 degrees for the stack effect to work effectively. Most portal frame roofs are 15 degrees.

5. Number, weight and metabolic rate of animals housed.
   This determines the heat and moisture the animals produce, and the effectiveness of the stack effect. Paradoxically, a very low stocking density can lead to poorer air-flow rates.

6. Prevailing wind/aspect of the building.
   Ideally, livestock housing should be sited facing away from the prevailing wind – to reduce the risk of draughts into the building and to carry away stale air. Open-sided buildings should face south so that stock get the maximum benefit from the sun. The proximity of buildings, trees, silage pits, and so on, may interfere with free flow of air.

7. Partitions across the flow of air.
   These can redirect air, sometimes causing re-circulation of stale air over stock. Solid partitions parallel with the direction of air flow minimize cross-flow of air between neighbouring pens and reduce spread of infectious organisms between neighbouring groups of cattle, but it is generally better to house animals in separate buildings to prevent contamination. Do not site young stock downwind of older animals that may be diseased.

If ventilation is found to be inadequate, it will usually be because of an inadequate ridge opening (which will not let stale air out quickly enough). Means of improving ventilation include increasing the inlet openings by installing space-boarding above a height of 1.5–2m, and removal of flaps that may have been installed to reduce draughts. Adult cattle generally do not require protection from draughts, and calves can be sheltered by a board at a level above the height of their backs. Slots can be created in the roof and solid gates can be replaced by slatted gates.

Fans can also be installed to create a forced ventilation system to increase air flow through large cattle buildings. Fans are used to create a presure gradient, either by sucking air out of the building to create negative pressure, or to blow air in to create positive pressure so

## Is Air Flow Through a Building Adequate?

| Question | Interpretation of answers |
|---|---|
| What is the incidence of pneumonia? | If pneumonia is a problem, building ventilation is likely to be the main factor, or at least a contributing one. |
| Does the air smell of ammonia? | There should be no smell of ammonia; if there is it usually indicates poor ventilation (also inadequate drainage and mucking out). |
| Is the air damp? | Damp conditions indicate inadequate ventilation. (Drainage should also be checked.) |
| Can condensation marks be seen? | Condensation is another indicator of damp conditions. |
| Are there a lot of cobwebs in the roof? | If so, air flow is probably not adequate. |
| Does smoke hang in the building? | This can be tested using smoke bombs. If air flow is good, smoke leaves building directly. |

natural air flow out of the building increases. Alternatively, fans can be used both to draw air in and suck air out (neutral pressure). It is usually best to blow air into cattle buildings. The physics of internal air movement using fans is beyond the scope of this book, and installation of these systems to improve ventilation should be undertaken with expert assistance.

## Flooring

Flooring of areas where cattle move and feed should be designed to minimize the risk of injury and disease: they should be non-slip, with no steps or steep gradients, and with no pooled slurry that may harbour infectious disease. Passageways should be of sufficient width to allow cattle plenty of space to pass one another, and there should be no blind ends or sharp corners. Bedded areas should be dry and comfortable, and provide enough space for stock to lie down whenever they want.

### Drainage

Effective drainage keeps the environment dry at floor level. This prevents build-up, in the bedding, of the infectious organisms that can lead to disease. A dry bed provides insulation, which is particularly important to the young calf, and it reduces soiling of cattle coats, which is essential for finishing beef cattle (beef cattle must be clean for slaughter – *see* Chapter 2). Effective drainage also reduces relative humidity (RH) and the survival of infectious organisms in the air, and it prevents pooling of slurry in passageways in dairy-cow housing (which is associated with increased risk of lameness).

Achieving good drainage is not always easy and depends mainly on the gradient or slope of the ground surface. Recommendations vary for different housing types and different classes of stock, with a slope of 1:20 providing drainage down a pen, and a slope of 1:33 for loafing areas, feed passages, and so on. The slope within a pen should be from back to front, with a drainage channel at the front (which must be kept clear of debris). There should be no drainage between neighbouring pens as this will spread disease. Water drinkers should be

*Cobwebs indicate that ventilation is not adequate. This house has no ridge, and air inlets are only via doors.*

placed away from bedded areas, so that beds do not get wet even if there is leakage. Drains should be the width of a brush so they can easily be kept clear, and should have solid covers to prevent injury.

### Floor Types

Floors may be concrete, hardcore, or slatted. Each has its advantages and disadvantages.

- Concrete. A concrete floor should be sloped at 1:20–1:33 to provide suitable drainage. Steep concrete slopes should not be laid because they become slippery when wet and may cause injury. The surface can be grooved after laying (ridged or hatched) so that it is non-slip, or skimmed smooth and flat for easy cleaning (for example in the milking parlour). It can be pressure-washed

*Grooved concrete provides a non-slip floor.*

and disinfected to eliminate infectious organisms. However, concrete becomes worn or cracked over time, particularly in areas where animals or farm machinery pass frequently. Cracks may cause injury and slurry may pool in worn areas, increasing hoof damage and risk of lameness. Potholes should be filled on a regular basis. The surface of worn concrete is often very smooth, and cattle may slip and be injured, so it can be re-grooved.

• Hardcore. Hardcore floors provide a suitable base for straw yards, provided the underlying soil is free-draining, although stones can get trapped in machinery used for cleaning out. Hardcore should be rolled/compacted to create a level surface.

• Slats. Slats must be suitable for the age and class of stock to be housed, providing enough drainage to ensure the floor remains dry, but with gaps that are narrow enough to prevent injury if youngstock are housed. Fully slatted floors are not to be used for breeding cows or replacement heifers (DEFRA Code of Recommendations for Cattle) and are also not suitable for calves less than twelve weeks old.

Cattle tend to get dirtier on slats than on straw (though a well-managed slatted yard is better than a poorly managed straw yard.) To keep cattle as clean as possible, the stocking rate on slats needs to be appropriate for

the age and weight of cattle (*see* Table 26, page 146). Understocking and overstocking tend to allow build-up of dung on slats, so as cattle grow they should be given progressively more space to maintain optimal stocking density throughout the rearing period.

Slurry storage and removal frequency must be adequate to avoid blockage of slats.

Water bowls are more suitable than water troughs over slatted floors because they provide no obstruction that will allow collection of dung.

*Bedding*

The purpose of bedding is to provide a non-slip, dry and comfortable bed with some thermal insulation at floor level. These various purposes may be served by different materials, but they must also be cost effective and suitable for use with available labour and machinery as well as the slurry-handling system. In cubicle housing, soft mats and mattresses can provide a comfortable substrate under a small amount of bedding. Suitable materials for bedding are straw, shavings and sawdust, paper, sand, and large wood chippings.

• Straw. Straw can fulfil all the roles itemized above. It must be used generously to be effective – a thick bed (60mm) of dry straw over concrete provides highly effective thermal insulation, while wet straw or

*Sand bedding must be at least 10cm deep. Its advantages are that it prevents growth of most bacteria, it is non-slip, it provides a comfortable bed, and it cushions the cow's feet and knees as she lies down and rises.*

a thin (12mm) bed provides little insulation and does not inhibit bacterial growth. A thin straw bed also provides little protection against injury in the cow cubicle. Straw can be costly, especially in areas that support little arable production, though the difference in cost between a good straw bed and a poor one is certainly less than the cost of extra cases of calf scour or environmental mastitis or a cow culled due to a swollen hock. Barley straw is generally considered better bedding than wheat. Straw should have a dry matter over 90 per cent, and it should be stored under cover or it becomes damp and may become contaminated with fungal growth. Automatic straw choppers allow straw yards to be bedded very rapidly.

- Shavings and sawdust. These are generally used on cubicle beds, and may be used over mattresses. Small particle size is associated with high bacterial growth and increased tendency to cling to udder and teat skin. Poorly stored shavings have been associated with environmental mastitis, in particular caused by *Klebsiella* species.
- Paper. Shredded paper can by used for loose-yarded cattle. It must be well managed or it becomes consolidated in some parts of the yard and boggy in others.
- Sand. Sand is considered by many to be the ideal bedding for adult cattle. It is inert, so it does not support bacterial growth, and it is associated with reduced incidence of clinical mastitis. It provides good cow comfort (if it is at least 10cm deep) because its particles move, conforming to – rather than compacting under – the animal's body. Sand disperses the impact around the feet as the animal rises and lies down, so it is particularly good for lame cows. It is associated with increased lying times. It is also non-slip, so it provides an ideal substrate for calving pens. Slurry handling systems may not be suited to use of sand bedding because it settles to the bottom of slurry lagoons and can cause wear to machinery. Sand gives little thermal insulation, so it is unsuitable for bedding young calves.
- Large wood chippings. These have been successfully used in roofless pens. The woodchip bed works like a septic tank, breaking down dung and urine through microbial action. The woodchip bed should be 30–40cm thick, and stocking density should be kept low.

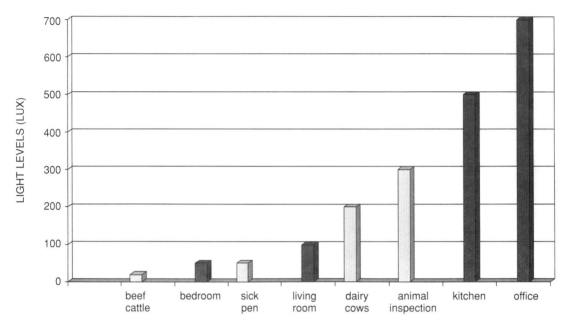

*Advised light levels in cattle housing compared to those recommended for the domestic and office environment.*

*Generous trough space is essential to encourage feed and water intake in dairy cows.*

## Lighting

Lighting level should be sufficient to enable the stockperson to observe cattle thoroughly (for example to look for signs of health and disease and to spot the bulling cow), and to carry out tasks such as cleaning out and feeding. The cattle require enough light to find feed, water and beds, and to avoid obstacles.

To enhance cost-effectiveness, the best possible use of natural lighting should be made. In smaller buildings (such as the monopitch calf-house), there will be natural light from the open front of the building, but in wide-span housing it is recommended that a minimum 10 per cent of the roof area should be left as roof lights, and preferably 15 to 20 per cent for walkways and handling areas. The level of artificial light should be bright enough for individual cattle to be identified. Long day lengths (sixteen to eighteen hours) – either natural or created through the use of artificial light – have been shown to increase both milk yield and growth rate in cattle, provided there is a period of short day length in the dry period.

## Access to Feed and Water

Feed access should be generous (*see* Table 26, page 146) and all cattle, including young calves, should have continuous access to fresh water. Any restriction of water intake may reduce feed intake and productivity. The sick animal is much less likely to survive if it suffers even mild dehydration.

## Building Layout

Buildings should be laid out for easy management and have good access to power supply, water, drains and roads (for transport of animals, feed, milk, and so on). For biosecurity reasons it is best if vehicles do not enter the farm premises, so buildings should be sited as close as possible to the road. There should be adequate access for farm machinery and sufficient flexibility to allow for expansion. Buildings should not be sited close to water courses because of the risk of pollution; and damp, low-lying sites, which will create ventilation

problems, should be avoided. Monopitch buildings should face south.

## Grouping

As with so many factors in cattle management, the group size of housed cattle is a compromise between various, often conflicting, factors. The needs of stockpersons and machinery must be balanced against the needs of the stock for production, behavioural welfare and disease control, all within the constraints imposed by available buildings and other farm enterprises. For ease of management and cost-effective use of large farm machinery, group size should be as large as possible, while small groups or individual pens are ideal for disease control.

Cattle are herd animals, and their behavioural needs are best met in stable groups. Bulls should be reared in groups of not more than twenty. Where animals must be moved from one group to another, it is best if several animals are moved at once (for example, dairy cows moving from the high-yielding to the middle-yielding group) to minimize stress. Movement between groups should always be minimized, particularly with growing beef animals and bull beef. All cattle, including young calves in individual pens, should be within sight and sound of other cattle.

For disease-control purposes, young cattle of different ages should be housed in separate pens and, preferably, in separate air spaces. Not more than forty calves should be housed in one air space, and an all-in, all-out policy should be adopted (the animals in one group being put in, and vacated from, the building at once, rather than adding extra animals or removing individuals from an established group).

# HOUSING FOR DIFFERENT CLASSES OF STOCK

## The Calf to Eight Weeks

The calf should be housed in a dry, draught-free environment, with minimum mixing of air between groups of different ages, and good lighting for observation of stock.

*Respiratory infection is likely where calves of different ages share one air space.*

The calf may be kept in an individual pen or in a group. Very large groups of young calves are not ideal because it is difficult for the stockperson to observe individuals adequately.

Where individual pens are used, the floor should be concrete – with a slope of 1:20 from back to front – and there should be a brush-width drainage channel in front of the pen with a slope of approximately 1:40. Dimensions of the pen are generally about 2m × 1m. Pen divisions may be solid or railed and should be easy to clean. Wooden divisions can be used but these are difficult to disinfect and may harbour infectious organisms. Individual buckets for milk, calf concentrate and water should be provided over the drainage channel outside the pen (to prevent spillage and for ease of management).

Calf-hutches (*see* page 60) can provide excellent accommodation for young calves, provided they are sited in a well-drained area. The calf must be offered a pen outside the hutch and

straw over the backs
of pens protect
calves from draughts

hurdles form the
front of each pen,
facilitating easy
observation of calves

large bales of straw
used to build pen
divisions

*An open shed provides flexible housing for young calves to weaning.*

*Weaned calves of similar
age penned in small
groups.*

## Individual Penning Versus Group Housing

| Individual Penning | Group Housing |
|---|---|
| Easy to monitor intake of milk and concentrate feed. | Individual monitoring impossible with demand feeding systems. |
| Easy to monitor signs of health/disease. | Difficult to monitor signs of health/disease. |
| Calf receives its share of milk – can drink in its own time. | Small or slow-feeding calf may be disadvantaged. |
| In principle, less disease spread, especially where partitions between pens are solid, but must be able to see other stock. | In practice there is little difference – quality of stockmanship is the key factor. |
| Higher costs for installation. | Lower costs for installation. |
| More work (feeding, cleaning out, dismantling pens between groups). | Less work. |
| In case of sickness, calf already isolated and can be nursed in its own pen. | Essential that isolation facilities are available for nursing sick calves. |
| Cross-suckling (where calves suck each other's navels) is not possible. | Cross-suckling may be a problem. |
| Calves cannot choose where to lie. | Calves can choose where to lie down (for example to escape draughts. |

*Calf shed built using existing stone walls for improved ventilation.*

air outlet

air inlet

air inlet

separate front gate of space-boarding for each pen

old stone wall

plenty of good-quality hay or straw to stimulate rumen

deep straw bed

must not be tethered. Bedding must be kept clean and dry. These require higher labour inputs than other systems, but provide useful isolation facilities away from the main housing.

For group housing, pen divisions should be easily dismantled for cleaning and should provide the flexibility to accommodate different groups.

Mixing age groups of calves increases the risk of disease, particularly pneumonia and scouring. Calves of different ages may be located in the same building provided that each age group has its own pen, and air space and drainage are kept separate, with solid walls between pens to a height of 2m and no cross-ventilation. Ideally, all the calves in a group are batched together ('all-in, all-out'). This is easy where calves of similar age are purchased from market in a batch of ten or twenty. It is more difficult with home-bred calves, where flexible grouping to suit calving pattern is best, with no more than seven days' difference between the youngest and oldest in one group. Optimal group size will depend upon the cattle enter-

prise: the large, block-calving dairy herd or beef-rearing unit may be able to house twenty or more calves of the same age in one group, while the small dairy unit calving throughout the year should provide for groups of two to four calves of one age. Pens should be thoroughly cleaned between groups. Walls that are impervious (such as coated concrete or coated breeze block) are much more easily cleaned and dry more quickly than wood walls do.

Calf feed should be prepared in a separate air space to the calves, because the use of large volumes of running water to mix milk powder and to clean equipment increases relative humidity (RH).

## The Weaned Calf

Once the calf is weaned, good ventilation is required to minimize risk of pneumonia. Calves are generally bedded on straw in a deep-litter system, but bedding should not be allowed to become sodden. Slatted flooring may be used, but not for calves under 250kg (unless straw bedding is used over the slats). Monopitch

corrugated iron sheeting baffle to protect calves from weather

corrugated iron sheeting roof

old stone wall

*Detail of air outlet at back of sloped roof.*

three sides solid to 1.4m to
protect animals from weather

open yard for loafing and feeding

monopitch roof over enclosed
area deeply bedded with straw

tractor access for feeding
using mixer wagon

high roof; open ended for excellent ventilation.
Draughts no problem for older heifers

space-boarding

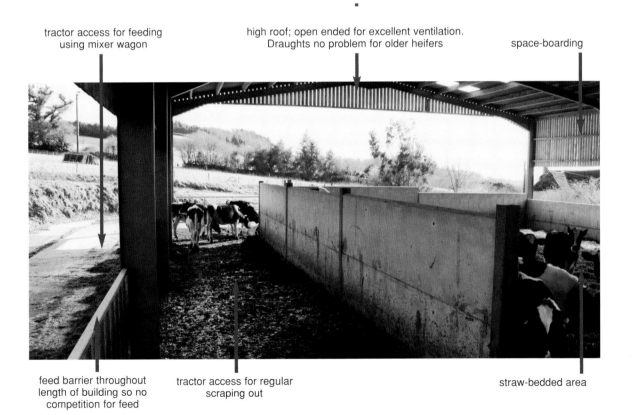

feed barrier throughout
length of building so no
competition for feed

tractor access for regular
scraping out

straw-bedded area

*Housing for weaned heifer calves should provide shelter and keep groups of the same age together. The house in the top picture was cheap to install, and flexible, but it offers the stockperson no protection from the weather. The bottom photograph shows a house in a heifer-rearing unit where large numbers justify the capital outlay.*

*Beef animals loose-housed in a straw yard with large air space and good ventilation provided by space-boarding and open sides.*

houses are commonly used, often with an open feed and loafing area at the front.

After weaning, calves should move to a larger house with more air space. Ideally they remain with animals of similar age, and group size does not increase to more than forty (*see* Table 26, page 146). It is important that feed intake is maintained, so trough space allowances should be generous to make sure that none suffers a growth check as a result of competition.

## Beef

The main purposes of housing older cattle are to prevent poaching of land and to ease management; in drier areas, roofless units are successfully used to rear beef cattle. In general, beef cattle are grown on and finished in housing that provides a lying area (either straw or slats) and an area for feeding and loafing. Trough space depends on whether cattle are fed ad lib or given a restricted ration (*see* Table 26).

### Bull Beef

Bull beef animals are rapidly reared to reach heavy weights for finishing at fourteen to fifteen months. There tends to be a lot of boisterous interaction between them, so feed barriers and partitions between pens must be robust. Handling facilities must be available within easy access to the bull beef pens, with minimal requirement for stockpersons to go into the pen to separate an individual (e.g. for treatment). Concentrate feed must be continuously available with plenty of trough space to ensure free access and therefore maximal growth rates, and there must be adequate clean water supply. Group size should be not more than twenty animals and, once the group is established,

---

### Potential Advantages and Risks of Cubicles

**Potential advantages**
- Dry bed.
- Undisturbed by other cattle.
- Increased stocking density.
- Reduced use of straw.
- Reduced incidence of environmental mastitis.
- Reduced labour required.
- Slurry easier to manage than solid dung.

**Risks**
- May be reduced lying time leading to reduced time cudding (and reduction in rumen function).
- Lameness.
- Injury.
- Increased incidence of cubicle rejection by cow, leading to:
  - Dirty cows.
  - Increased risk of environmental mastitis.
  - Increased culling.

---

new animals should not be introduced or serious fighting and injuries may occur. Female cattle should be kept well away from bull-beef rearing pens.

## The Suckler Herd

In spring-calving herds, calves are generally weaned off before they are housed, but housing for the autumn-calving suckler herd must provide for the various needs of both adults and calves. Water troughs must be at a height that calves can reach, and there should be a creep area to enable calves to feed and rest away from adult cattle.

It is usually preferable for suckler cows to calve at grass rather than during the housed period, but calving boxes should be provided if cows are calving indoors.

## The Dairy Cow

Although practical issues such as ease of management and costs must be considered in housing the dairy cow, it is cow comfort that is of paramount importance. There are two reasons for this. The first is that the metabolic demands on the high-yielding cow are heavy, and unless she spends most of her time eating or chewing the cud she cannot meet these metabolic demands and the result will be disappointing yields and/or metabolic disease. The cow likes to lie down to chew the cud, and she needs a comfortable bed. The second reason why cow comfort is so important is to prevent lameness. Lameness is extremely common in

dairy cows – with average UK incidence about 25 per cent – and it is a contributory factor in poor fertility and a major reason for culling in many herds. Research has shown that much of the damage to the foot that predisposes to lameness is the result of poor housing conditions and reduced lying time in the period around calving. The first-calving heifer is particularly vulnerable, and providing a comfortable bed can help to reduce the stresses on foot tissues that lead to permanent damage.

As well as a comfortable bed, the cow should be able to stand up and lie down easily. Cows tend to lie down for only about one hour at a time, and at pasture they get up and lie down at least ten times a day, with total lying time of about eleven hours.

### Cubicles

When cubicles and cow kennels were introduced to the UK in the late 1960s, they offered a cost-effective option for providing the cow with a dry bed at a higher stocking density and with lower straw use than that required for a straw yard. While these potential advantages remain, in practice the cubicle house may also be associated with problems. In order to maximize potential advantages and minimize risks:

1. The cubicles should be well designed. The ideal cow cubicle provides minimal restriction to the cow as she lies down, while she remains recumbent and as she rises. Partitions between cubicles should

| Table 28: Cubicle lengths according to weight | | |
|---|---|---|
| *Weight of cow (kg)* | *Total length of bed (m) (open front)* | *Total length of bed (m) (closed front)* |
| 500 | 2.05 | 2.05 |
| 600 | 2.15 | 2.15 |
| 700 | 2.3 | 2.55 |
| 800 | 2.4 | 2.75 |

(From *Housing the 21st Century Cow*, MDC 2006.)

be robust and without sharp projections that may injure the cow. The cubicle should have a slope of approximately 1:20, from front to back, to allow drainage. If a kerbstone is provided, it should be rounded to minimize hock injury and should not prevent drainage of leaked milk, urine or dung.

2. The cubicles should be the correct length and width (*see* Table 28). Length should be sufficient for the cow to lie in comfort and lunge forward to rise, whilst minimizing soiling of the back of the cubicle with dung or urine. (The brisket board also helps to position the cow correctly on the bed.) Width should allow even the

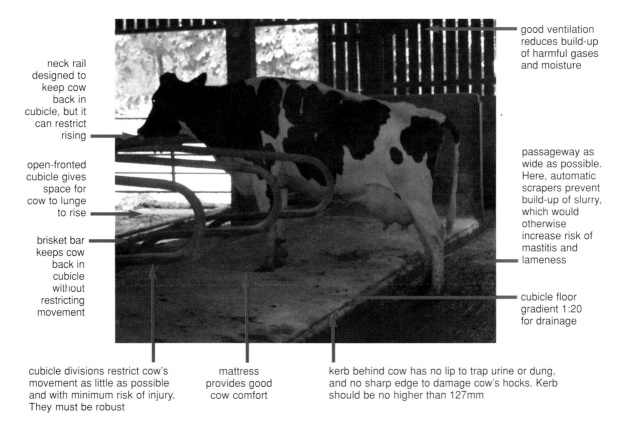

neck rail designed to keep cow back in cubicle, but it can restrict rising

open-fronted cubicle gives space for cow to lunge to rise

brisket bar keeps cow back in cubicle without restricting movement

good ventilation reduces build-up of harmful gases and moisture

passageway as wide as possible. Here, automatic scrapers prevent build-up of slurry, which would otherwise increase risk of mastitis and lameness

cubicle floor gradient 1:20 for drainage

cubicle divisions restrict cow's movement as little as possible and with minimum risk of injury. They must be robust

mattress provides good cow comfort

kerb behind cow has no lip to trap urine or dung, and no sharp edge to damage cow's hocks. Kerb should be no higher than 127mm

*Key features of the cow cubicle.*

---

**Cubicles: Common Faults and Observations**

**Cubicles too short and/or too narrow**
- Cows stand with hind feet in passageway.
- Difficulty lying down and rising.
- High rejection rate of cubicles (especially big cows).
- Cows choose cubicles away from outer walls (inner rows have more lungeing space).
- Hock injuries (cubicle too short); rubbed pin bones and shoulders (cubicle too narrow).
- Rubbed rails in cubicles.
- Lameness.

**Passageway too narrow**
- Difficulty backing out of cubicles.
- Lameness.
- High rejection rate of cubicles (especially heifers).

**Head rail too low or placed too far back**
- Cows stand with hind feet in passageway.
- Difficulty lying down and rising.

- Rubbed/broken neck rails.
- Injury over necks.
- Dimensions inadequate.

**Poor design of partitions**
- Rubbing of partition rails.
- Cows stand with hind feet in passageway.
- Difficulty lying down and rising.
- Injury.

**Poor cubicle management**
- Dirty beds.
- Dirty cows.
- High incidence of environmental mastitis.
- Inadequate bedding.

**Inadequate passageway scraping and/or flat passes that create slurry pooling**
- Slurry in passageways
- Dirty beds
- Dirty cows
- High incidence of environmental mastitis and lameness

---

largest cow to lie comfortably without damage to her hips and shoulders, but not so wide that small cows are tempted to lie diagonally or backwards. Optimal cubicle width is 115cm.

3. There should be at least one cubicle per cow. All cows like to lie down at the same time, and having plenty of available cubicles ensures this.

4. Cubicle beds should be comfortable, cost effective and prevent build-up of environmental pathogens that may cause mastitis. Cow mats increase lying times, have been found to be better than straw or sawdust for preventing injury to hocks, and reduce bedding requirements. Sand does not allow growth of mastitis bacteria, but generally cannot be used with automatic scrapers and must be at least 10cm deep to provide adequate comfort. Chopped straw is retained better on the bed than long straw.

5. The house should be well ventilated but protected from draughts and rain. Poor ventilation or rain wets cubicle bedding and facilitates growth of harmful bacteria. Cows tend to reject draughty cubicles where the rain blows in, and it is the most vulnerable animals, such as first-calved heifers, who are excluded from the more comfortable beds.

6. The cubicle house should be laid out to allow easy movement of animals and free access to feed and water, and to facilitate cleaning out. Passageways must be wide enough (3m) for cows to back out of cubicles and turn without difficulty, and pass each other without confrontations. A cross-passage should be provided every twenty cubicles to help prevent bullying.

7. The cubicles and passageways must be kept clean. Scrupulous attention to detail is vital. Dung pats and soiled bedding should be removed – and passageways scraped –

## Identifying Common Problems in the Cubicle House

### Observation of the cows
- Is there damage to hocks, knees, hips and necks?
- What is the prevalence of lameness during the housing period (also other causes)?
- How often do cubicle injuries such as damaged teats and cows stuck in cubicles occur?
- Are there dirty cows (owing to cubicle rejection and/or dirty cubicle beds)?
- Are cows' feet dirty?
- Which cows are most at risk (e.g. biggest cows, first-calved heifers)?

### Disease records
- Mastitis during the housed period (also other causes).
- Lameness during the housed period (also other causes).

### Cow behaviour
- Cows should be observed 1–1.5 hours after milking. How many are not lying down?

- Cow preferences – are some cubicles not being used?
- Are there cows lying in passageways or in the loafing yard?
- Do cows stand in cubicles with back feet in passageways? (This may indicate lameness.)
- How easily do cows lie down and get up? Are they reluctant to go into cubicles or lie down?
- Are cows correctly positioned when lying in the cubicle?

### Cubicle dimensions and structure
- Is there evidence of rubbing, e.g. on rails?
- Are there any broken cubicle rails?
- Is there a kerb lip?
- Is the bedding clean and are beds draining?
- Is there enough bedding to provide adequate comfort?
- Are there enough cubicles (for number of housed cows)?
- What are the dimensions? Are all cubicles the same width or is there variation?

*Poor cubicle housing increases the risk of lameness.*

cubicle too short, so cows do not lie in

inadequate bedding so cows are not comfortable. They therefore spend more time standing and there is increased risk of swollen hocks and knees

narrow passage, which fills with slurry unless frequently scraped. Narrowness also means cows must twist as they back out, adding stress to the feet

hind feet in passageway causes strain on tendons, and feet are bathed in slurry

*Pooling of slurry around automatic scraper close to water trough increases the risk of lameness caused by infectious organisms.*

at least twice a day. Automatic scrapers should be set to clean the passageways every two hours during the day – but not at night (to avoid disturbing the cows).

A great variety of cubicle designs and cubicle-house layouts have been used with a similarly great variety of success and failure; you need to look at how a building is performing

*You can estimate cow comfort by observing how cows lie in their cubicles.*

*Old-style cubicles can work well provided they are well bedded and not too crowded.*

*The ideal cow cubicle provides minimal restriction to the cow as she lies down, while she is recumbent, and as she rises.*

to determine how good it is. The checklists on pages 168 and 169 will help you to trouble-shoot cubicle problems.

*Tackling Cubicle Problems*

The cause(s) of many cubicle problems may be obvious, but finding practical solutions is often more difficult. Problems frequently occur because cubicles are old, because herd size has increased and because cows in the herd are bigger. We usually need to look to improve existing facilities to provide more space for cows, and in some cases the cubicle house needs to be replaced. Cubicle modification is a specialized skill beyond the scope of this book, but generally the situation may be improved by reducing the number of cows in the house. It may be possible to separate first-calved heifers or the largest cows in the herd, giving the remainder more choice in the cubicle house. Extra bedding can improve cow comfort and increased frequency of scraping out and cleaning the backs of cubicle beds will keep the cubicle house cleaner. Frequent application of lime to the back of cubicles can help to prevent build-up of infectious organisms. In order to give the cows more lungeing space the head rail can be replaced with a brisket board or, in some buildings, the wall at the head of the cubicles can be removed. A kerb added to the back of the cubicle increases its length (although this should not be done if passageways are narrow).

*Straw yards provide good comfort for housed cows. For these beef cows, a straw-chopper blows in straw every other day, and the feed passage is scraped daily (scraping is more frequent for dairy cows because their output of urine and faeces is higher).*

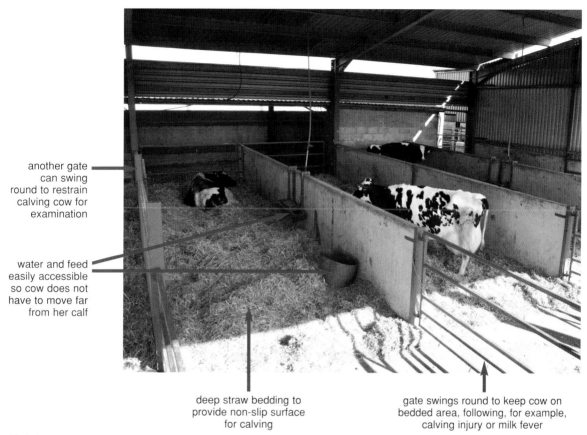

another gate can swing round to restrain calving cow for examination

water and feed easily accessible so cow does not have to move far from her calf

deep straw bedding to provide non-slip surface for calving

gate swings round to keep cow on bedded area, following, for example, calving injury or milk fever

*Calving pens.*

## The Straw Yard

The potential advantage of the straw yard is that cattle tend to spend more time lying down. This increases cudding time and efficiency of rumen function, which is particularly important in the high-yielding dairy cow. Lameness incidence is generally lower on straw yards, but, unless well managed, incidence of environmental mastitis tends to be higher.

The key to success with loose housing is to keep the bed dry. Drainage must be adequate, with a slope of 1:60 across the bedded area towards the feed passage. Flooring should be concrete or (if land is freely draining) hardcore. The straw yard must not be overstocked (remembering that stocking density is lower for high-yielding cows because they produce more dung and urine). Water troughs should be sited off the bedded area so that bedding does not get wet if the troughs leak and so that cows must drink while standing off the bedded area (to avoid poaching). The house should be well-ventilated. Plenty of dry barley straw bedding should be used, with twice daily bedding (a third in the morning and two-thirds at night). On average, 2 tonnes of straw per cow will be required for a 180-day housed period.

Feed passages should be scraped twice daily. The straw yard must be completely cleaned out every three to six weeks – the optimal frequency depending upon factors such as stocking rate, ventilation, environmental temperature and humidity, and diet (which determines consistency of dung). Any delay in cleaning out the straw yard leads to increased risk of environmental mastitis. Fencing is generally used to confine cattle to the feeding area while the yard is cleaned out.

Group size in a straw yard should ideally be up to forty animals and certainly not more than sixty.

## The Calving Cow

The cow prefers to calve in isolation or at least well away from other cattle. In many herds the cows calve in the dry cow yard, but this is not ideal. If she is unable to get on with calving without disturbance, there may be increased risk of calving difficulty, and after calving mismothering is likely. It may also be difficult to separate the cow for examination if she is calving with a group of other cattle.

Ideally, therefore, a calving box should be available to provide for the needs of the cow both during calving and for the period immediately after calving while the cow remains with her calf. It should also be suitable for the accommodation of any cow that suffers a calving injury and may need to remain in the box for a few days.

The calving box should be quiet to allow the cow to calve with minimal disturbance. However, there should be easy access for the stockperson so that the cow can readily be monitored. A non-slip floor reduces the risk of injury during calving. This can be either a fresh, deep-litter straw bed (which must be well drained and dry to prevent build-up of infectious organisms) or straw over a sand bed. Clean straw over concrete is not ideal because the cow may slip. The calving box should provide a space allowance of 16m², or 12m² if calving in small groups, with no projections that may cause injury. Good lighting (including 20 per cent natural roof lights) is important, and there should be handling facilities to enable calving assistance to be given; most simply this can be a gate that is pulled around the cow and fixed with a rope. If the cow chooses to lie down, the gate can be released without difficulty.

A supply of fresh water is essential since the cow may not have drunk during parturition, which may have lasted some hours, and water requirements increase very rapidly after calving. It is also important that the cow can easily be fed with an appropriate ration; appetite is low for the twenty-four to forty-eight hours around calving and we want to encourage maximum feed intake as soon as possible after the calf is delivered. In dairy herds the calving boxes should be sited close to the milking parlour; in some herds a vacuum line is run to the calving boxes so that freshly calved animals (at high risk of 'doing the splits' on slippery concrete) can be milked in the box.

Sufficient calving boxes are needed to accommodate all calving animals, allowing for any that might require an extended stay (for instance after a difficult calving), and to allow time for cleaning out and addition of fresh bedding.

## Sick and Quarantined Animals

An isolation unit is essential in order that a sick animal can be taken out of its group – both to prevent spread of disease and to allow for extra nursing care, individual feeding and close monitoring. Purchased stock can also be quarantined in the isolation unit. Air space and drainage should be separate from other housing, but if it may be used to house dairy cows it must be sited within easy reach of the parlour. Walls and floor should be solid to facilitate disinfection between each individual's occupation, and a deep, dry, straw bed provided. The facility should be flexible, with gates or hurdles so that it can be used for different classes of stock. It needs good lighting and handling facilities for potential veterinary interventions. A heat lamp is valuable for the care of sick calves.

# HANDLING FACILITIES

Cattle must be handled from time to time, both for routine herd procedures, such as vaccination or weighing, and for examination and treatment of the individual animal. Suitable handling facilities are therefore essential in the management of every herd. Stress levels and risk of injury (both to humans and animals) can be minimized – and speed and efficiency can be increased – if

## Cattle Behaviour and Handling

| Cattle Behavioural Trait | Consequences for Handling |
|---|---|
| Cattle like to keep in view of other animals at all times. | A holding race should have space for several animals. |
| They may panic and injure themselves or others if left alone. | Never leave one animal isolated. |
| Cattle will not readily approach a dead end. | They will proceed better along a gently curved race (radius 3.3–5m) than along a race with corners. |
| Cattle will move towards light (particularly daylight) but not towards a darkened area. | The race should lead out of a building rather than into it. |
| Cattle are easily distracted or frightened by anything unexpected around the handling facility. | Solid walls may seem ideal, but are not practical because they make it difficult to move the cattle forward or to handle them. |
| Cattle vision at ground level is poor, so they may baulk at shadows at ground level. | Avoid changes in texture or light at ground level. |
| If cattle escape once, they will try again. | Design should prevent the first escape: barriers and gates that are too high to jump over, too strong to be crashed through, and with no gaps to be squeezed through. Ideally the design offers cattle no choice but to proceed correctly. |
| It is much easier to move cattle forwards than backwards. | Use a tightly packed chute to funnel cattle ahead and do not let them turn round. This is difficult where animals of different sizes are to be handled through the same race. |
| Cattle move much more predictably at the walk than once they are running. | Make sure that you handle them quietly to prevent a stampede. |
| Loud or unpredictable noises startle cattle. | Do not shout. Keep barking dogs, etc., away. |
| Cattle remember frightening and painful experiences for many months. | Avoid rough handling and use of electric prodders or sticks. Handle cattle occasionally without doing anything aversive: simply weigh them or apply a pour-on treatment to combat flies or worms. |
| Once agitated or confused, cattle behave unpredictably and are likely to injure themselves or others. | Handle cattle calmly and quietly. |
| Cattle follow the lead of a dominant animal. | Keeping a quiet older animal with boisterous youngsters may make handling much easier. |

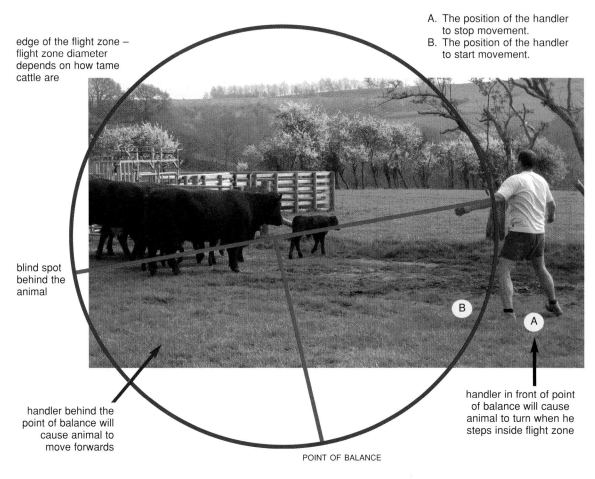

A. The position of the handler to stop movement.
B. The position of the handler to start movement.

edge of the flight zone – flight zone diameter depends on how tame cattle are

blind spot behind the animal

handler behind the point of balance will cause animal to move forwards

POINT OF BALANCE

handler in front of point of balance will cause animal to turn when he steps inside flight zone

*The flight zone and point of balance for handling cattle. (Based on Grandin's behavioural principles of livestock handling.)*

handling facilities are well designed. Some principles of cattle behaviour need to be taken into account to understand how cattle can be moved successfully through the facility (*see* box on page 175).

## The Flight Zone

The 'flight zone' is the distance within which cattle will move away if a human approaches (*see* illustration above). The radius of the flight zone depends upon how wild or tame the animal is, which in turn depends upon factors such as its breed and its experience with humans. Cattle that are gently and frequently handled (such as most dairy cows) may have

a flight zone of less than 1m, while outdoor, extensively reared beef cattle may have a flight zone of 8–50m.

Cattle generally move predictably if you work at the edge of the flight zone; if you step outside the flight zone, the animal stops, if you step into the flight zone – towards the animal – it moves off. If you step deep into the flight zone, the animal tends to bolt away or to turn back and rampage past you. This makes it difficult and dangerous to move wild cattle in a confined area (where you are certain to be well inside the flight zone); it is better to stay outside of the flight zone and move cattle by opening and closing gates.

The direction of cattle movement depends on whether you are in front of, or behind, the so-called 'point of balance' (*see* illustration opposite). To move cattle forward, walk at an angle behind the point of balance. To turn them, step forward across the point of balance.

## Safety

The crush should immobilize the animal sufficiently to allow the stockperson to work with minimal risk; for example, the foot should be tied to a block for trimming and a halter or bulldogs should be used if the head must be handled. Ideally the crush should have doors that can be opened to examine and/or treat the udder, the head, or a foot, without the risk of a hand or arm being injured between the crush and the animal. A bar behind the animal should be long enough that it cannot be dislodged and fly out if the animal struggles or

*Excellent handling facilities are essential for horned cattle for human and cattle safety.*

good lighting

direct access from milking parlour

gate operated from parlour to direct cows into race

slats to drain floor after hosing down

race for two to three cows

foot crush

after treatment cows can be let back with the herd or kept in a holding pen

*Good handling facilities close to the parlour simplify lameness treatment.*

*The footbath is an essential facility because footbathing is an integral part of lameness control in the dairy herd. A range of different treatment solutions can be used in the footbath, which should be chosen according to the foot disease to be controlled. The footbath should be easily accessible from the parlour and include a wash bath to clean the feet before the cows enter the footbath. The treatment bath must be at least 3m long.*

kicks. The crush should be kept in good repair so it restrains the animals effectively. A gate should be sited between the crush and the race so animals are kept back if a stockperson has to work behind the animal in the crush.

Handling facilities should enable the stockperson to divide cattle into small groups, fairly tightly packed, rather than having to use a large pen where they can charge about. No one should work in the race with cattle that are not restrained.

Handling facilities should be sited near to pens so that the cattle do not have an opportunity to break out or get excited before they enter the yard. For dairy cattle the crush should be adjacent to the parlour so that individuals requiring treatment can be separated after milking.

## Artificial Insemination Pens

Conception rates are likely to be reduced if cows are stressed prior to AI, so it is essential that AI facilities should be comfortable and quiet, and allow cows to relax. Since they sometimes have to wait for some hours, ideally they should be able to lie down, and there should be feed and water available. However, the AI pens should be designed to restrict cow movement sufficiently to facilitate easy insemination, and in dairy herds they should be sited near to the parlour so that cows to be inseminated can be separated after milking.

# CHAPTER 9

# Health and Disease

All disease is costly in both economic and welfare terms; at its worst it results in death, forced culling or permanent loss of production (for example, stunted growth, so that a beef animal can never successfully be finished, or loss of milk production due to chronic mastitis), but even mild disease causes a minor dip in production or fertility. There are also the costs of diagnosis and therapy and the extra labour that is required to treat sick animals.

In an ideal world, therefore, there would be no disease. However, disease does occur, whatever the size of herd and whatever the level of husbandry skill or management investment. Disease costs can be minimized by introducing preventive measures that are cost-effective and by tackling any emerging disease thoroughly and effectively.

This chapter explores the ways in which this can be done. Signs of health and disease are also described.

*This scouring calf collapsed. She recovered after intravenous fluid therapy to correct dehydration and acidosis.*

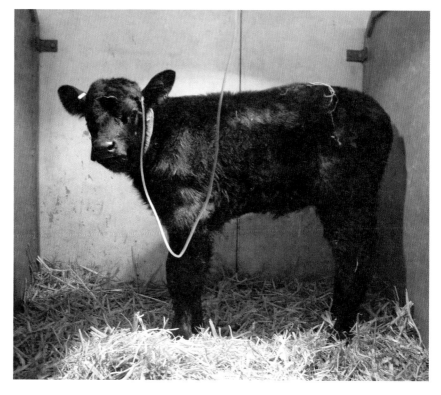

---

## Infectious and Non-Infectious Diseases

### Infectious diseases
These are caused by a variety of pathogens:

- Viruses.
- Bacteria.
- Fungi.
- Parasites.

Many types of viruses, bacteria, fungi and parasites live on animals without causing any ill-effect. Pathogens are those that are able to invade and colonize the animal's tissues, to multiply and damage those tissues to cause disease.

### Non-infectious diseases
The causes of non-infectious disease are varied. The roots of such disease may be:

- Congenital (present at birth).
- Genetic.
- Metabolic.
- Nutrition-related.
- Toxic.
- Traumatic.
- Neoplastic (cancer).

---

# WHY DISEASE OCCURS

The various major categories of disease can be grouped as infectious and non-infectious diseases.

## Infectious Disease

At the heart of infectious disease there is a simple dynamic between the animal, the infectious agent and the animal's environment. Whether or not disease occurs when an animal meets an infectious agent depends on the animal's defence mechanisms and the infectious agent's numbers and pathogenicity (ability to invade and damage tissue). In some cases (such as salmonellae) very small numbers of pathogens can be enough to cause disease; in others, disease tends to occur only if the animal is overwhelmed by large numbers of pathogens (as in the case of coccidiosis). The environment may increase the likelihood of disease occurring, either by disruption of host defence mechanisms or by allowing survival and multiplication of pathogens so that the chance of infection increases (for example, a moist and dirty bed is likely to cause environmental mastitis in cows and scouring in calves).

The animal protects itself from infectious disease in a variety of ways. For example, the skin protects the body and if its integrity is breached by a cut or graze infectious agents may invade. The teat sphincter of the cow has a variety of anatomical mechanisms, including a muscle to contract the teat orifice after milking and a keratin lining that acts like tape, 'sticking' bacteria to it to stop them from penetrating the udder. In the reproductive tract the cervix remains tightly closed during pregnancy to protect the growing foetus from infectious agents outside.

Environmental factors may damage or disrupt these natural mechanisms. For example, the teat sphincter can be damaged by the milking machine, and high levels of ammonia in a

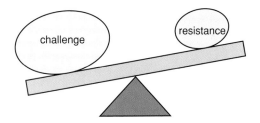 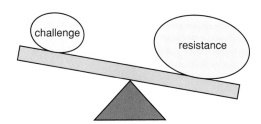

*Whether or not disease occurs depends upon the balance between challenge from infectious organisms and the resistance of the animal.*

dirty calf-house can damage the lining of the airways, allowing invasion by pathogens. The animal may also mount a specific immune response against a pathogen to which it is exposed. This results in a cellular and/or humoral and/or surface response:

- A cellular response is an increase in various types of white blood cells that are directed at defeating the infectious agents.
- A humoral response involves the production of circulating antibodies.
- Surface immunity by antibodies lining the gut, respiratory and reproductive tracts.

This cellular and humoral immunity may be lifelong or it may wane after only a few weeks. Once the 'memory' of a disease is lost, the animal becomes susceptible again. Immunity may be acquired naturally, when the animal is exposed to the disease-causing agents (pathogens), or artificially by vaccination.

*This cow has toxic mastitis after calving. The cow's defence mechanisms are low around calving, so she is more susceptible to infectious disease. Cows may die from toxic mastitis, despite veterinary treatment.*

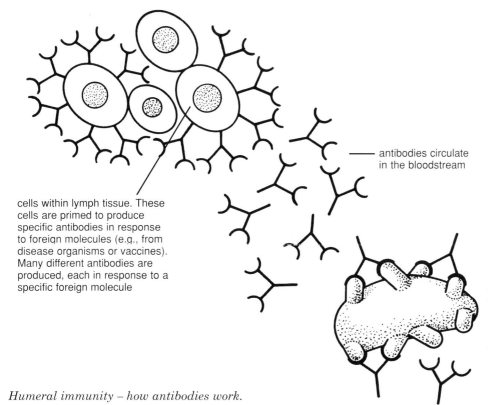

antibodies circulate in the bloodstream

cells within lymph tissue. These cells are primed to produce specific antibodies in response to foreign molecules (e.g., from disease organisms or vaccines). Many different antibodies are produced, each in response to a specific foreign molecule

If the antibodies meet their target foreign molecule, they bind and neutralize it. This in turn facilitates breakdown by white blood cells.

*Humeral immunity – how antibodies work.*

The ability of an animal to respond to a disease challenge may be compromised by a number of factors, such as:

- The presence of other disease, either infectious or non-infectious. Any disease is likely to place the animal under a degree of physiological stress. However, some infectious diseases, including bovine viral diarrhoea (BVD), have a direct depressant effect on the animal's immune system, making it more susceptible to attack from other pathogens.
- Adverse environmental conditions. These include conditions that increase exposure to infectious agents and can damage host defence mechanisms or cause stress and discomfort (for example, poor walkways/tracks or flooring that damages the foot, or poor milking-machine function).
- Provision of a ration that does not meet the animal's physiological requirements.
- Provision of a ration that is inappropriate for normal digestive function – for example, excessive feeding of fermentable carbohydrates, leading to rumen acidosis.

A natural physiological suppression of the immune system is associated with the period immediately around calving. This cannot be prevented, but to minimize its effects the calving cow should be in an environment where disease challenge is at a minimum.

## Non-Infectious Disease

Non-infectious diseases may be:

- Congenital.
- Genetic.
- Metabolic.
- Nutrition-related.
- Traumatic.
- Toxic.
- Neoplastic (cancer-forming).

Congenital diseases are those seen at birth. They may result from a variety of factors, ranging from spontaneous genetic mutations to specific chemical or infectious insults to the foetus during development.

Genetic diseases result from defects within the genetic code and are usually passed from the parents' genes (hereditary), though they

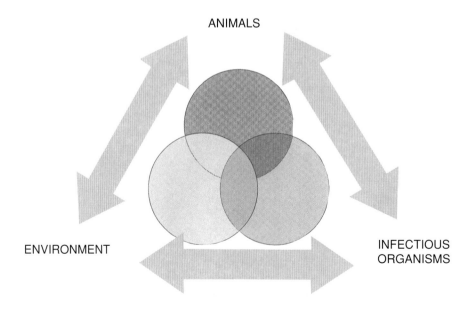

*Interaction of factors in the development of infectious disease.*

may occur as a result of spontaneous genetic mutation. A classic example of hereditary disease is complex vertebral malformation (CVM).

Metabolic diseases are an important source of loss in many herds. They occur when normal body function is disrupted, and are frequently associated with inadequate or incorrect nutrition, particularly around calving. They are far more common in dairy cows than in beef cows. In many herds, few animals show clinical signs, even though large numbers are affected sub-clinically (i.e., large numbers of affected animals show no clinical signs). The most commonly recognized metabolic disease is milk fever, caused by a fall in blood calcium levels after calving. Ketosis is a condition resulting from excessive energy deficit, and fatty liver is a complex disease characterized by excessive fat deposition in the liver resulting in disrupted function. Hypomagnesaemia, often seen in

beef animals in spring, is due to a fall in blood magnesium.

Nutrition-related diseases are common and various. They may occur as a result of inadequate supply of a particular essential nutrient, such as copper or cobalt, or toxicity from excess copper or selenium, or as a result of nutritional imbalance causing abnormal rumen function, such as acidosis. Many other multi-factorial diseases, including lameness, sub-optimal fertility and left displacement of the abomasum (LDA), are also associated with nutrition.

Traumatic diseases occur as a result of direct trauma. For example, chronic abrasion can cause hock lesions in cubicle-housed dairy cows.

Toxic disease in cattle is usually the result of the ingestion of toxins, including moulds that produce mycotoxins in the feed.

Neoplastic and immune-mediated diseases are not commonly encountered on cattle farms.

Initially, cow is wobbly when she walks; later she lies down. May attempt to rise, but is unable to take weight, especially on hind legs; may lie flat out. Temperature is normal or below normal. If affected before calving, she will not be able to calve until calcium balance is restored.

sometimes bloat, no rumen movements, no cudding

neck often curved round against flank

cold extremities

dull; not interested in surroundings. Does not eat or drink

nose dry

little or no dung passed

*Signs of milk fever. Milk fever usually occurs after calving, occasionally just before calving. It must be treated with calcium injected into the vein. (In mild cases, calcium is injected under the skin or given orally.) Check also for calving injury, blood loss after calving, toxic mastitis, and infection of the uterus.*

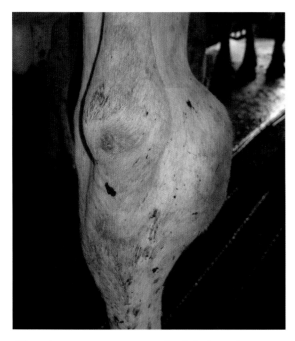

*Chronic trauma may cause hock lesions that are difficult or impossible to resolve.*

# DISEASE PREVENTION AND CONTROL

It has already been said that disease occurs as a result of the interaction between the animal, the environment and infectious or parasitic pathogens. Effective disease prevention and control requires the balance to be tipped in favour of the animal. The steps to be taken are:

1. Non-specific preventive measures that improve host (animal) defences against disease, improve the environment and reduce the spread of pathogens.

2. Monitoring of nutrition and production.

3. Early identification of disease in the individual and/or the herd.

4. Accurate and thorough investigation of disease as it occurs.

5. Treatment of sick animals.

6. Specific control measures to prevent further cases.

7. Eradication of disease.

8. Biosecurity.

## Non-Specific Preventive Measures

### The Animal

The animal's defences against disease are most effective if:

- Nutrition is excellent. This includes the supply of energy and protein and the provision of minerals and trace elements. Nutrition must be appropriate for the stage of growth and production (*see* Chapter 5).
- Calves receive plenty of good-quality colostrum in the first six hours after birth (*see* Chapter 3). Colostrum is rich in antibodies and provides the calf with all its immunity against infectious disease for the first weeks of life.
- Stress is reduced to a minimum. For example, herd groups should remain stable and cattle should be handled quietly and transported slowly and gently.
- Sub-clinical disease is controlled. Parasitic infestation, metabolic conditions (for example, milk fever and ketosis) and trace element deficiencies that are not severe enough to cause clinical signs may increase animals' susceptibility to other clinical disease. These sub-clinical conditions produce no clinical signs, so can usually be detected only through laboratory analysis of blood or dung samples. Ideally, samples should be taken for analysis from at-risk cattle: for example, cattle in their first year of turnout should have dung samples taken to check for parasites.
- Management of the transition cow (in the weeks before and after calving) is optimal. Dairy cows are at highest risk of metabolic and infectious disease in the first weeks after calving, and these risks can be reduced by managing them with extra care

*Colostrum intake within six hours of birth is vital to protect the calf's health.*

for the last three weeks of the dry period until after calving (*see* Chapter 5).

- Cattle have been genetically selected for suitability for the production system. For example, hardy native breeds that produce small, vigorous calves are best suited to extensive hill beef production, where weather conditions are harsh, feed quality is poor and there can be few inputs from the stockperson. Dairy farmers who have adopted the extensive grazing system, where cows are at grass for nine or ten months of the year, often with little supplementary feed, generally find that Shorthorn- or Channel Island-cross cows fare better than the Holstein.

- Routine management procedures are carried out. Examples include foot-trimming and footbathing. Lameness is common and increases susceptibility to a variety of other diseases, so lameness-control measures will improve overall herd health. Another example is proper milking-machine maintenance: improper milking-machine function will contribute to mastitis in a variety of ways.

### The Environment

Environmental factors may increase risk of disease either by reducing the animals' resistance or by increasing their exposure to pathogenic agents. The risk factors that contribute to lameness in dairy cattle are outlined

*Foot-trimming. Routine foot care helps to prevent lameness. (Photo: Kingfisher Veterinary Practice.)*

*Cows kept in dirty conditions are more prone to lameness and mastitis.*

in Chapter 8 and include those that alter hoof horn and wear, such as concrete quality, yard design, cubicle dimensions, stony or uneven tracks and walkways, as well as those that increase exposure to infectious organisms, including slurry-pooling, frequency of scraping out, and mud and muck around water troughs at pasture. Similarly, the risk of pneumonia in young cattle is increased where ventilation and drainage is poor, because there is a build-up of infectious organisms and of ammonia in the atmosphere. Disease is more likely where there are more pathogenic organisms and where ammonia reduces the effectiveness of the calf's natural defences against invasion. Draughty conditions cause chilling in very young calves, increasing susceptibility to disease.

*The Pathogen*
The risk of infectious or parasitic disease depends on the causative organisms' ability to invade and damage the animal's tissues (their pathogenicity), their number and the likelihood of spread.

The number of organisms present, and their chance to spread, depends in turn on:

- Hygiene. This is vital – for example, during feed preparation of milk for calves, and to reduce spread of contagious mastitis in the parlour by thorough application of post-milking teat disinfectant.
- Regular cleaning out and bedding up. This eliminates infectious organisms and keeps the floor environment dry, so that infectious organisms are less likely to survive or multiply. These measures are less effective if housing is poorly drained (*see* Chapter 8).
- Ventilation. Effective ventilation removes large numbers of potential pathogens, and dries the air to reduce droplet spread of infectious organisms.
- Isolation of sick animals. Sick animals may shed millions of infectious organisms that can infect healthy stock. Ideally, the infected animal is kept separate from the healthy group to reduce spread, for example by removing a scouring calf from the main pen; where there is more than one animal affected, for example by chronic mastitis, an 'infected herd' of cows can be created so that they can always be milked last. If the disease is highly contagious (for example, salmonellosis), it may be necessary for stockpersons to use a disinfectant foot-dip, separate overalls and disposable gloves to avoid spreading disease between animals or to themselves.

*An isolation pen limits spread of infectious disease within a group and allows individual care of the sick animal.*

- All-in, all-out housing policy. This allows pens to be thoroughly cleaned between groups and prevents the spread of disease from older to younger cattle that are added to a group.
- Closed herd or careful selection of stock for purchase according to disease status, to avoid buying in disease (*see* biosecurity section, page 192).
- Quarantine (ideally thirty days) of purchased stock.
- Elimination of vectors that carry disease, such as snails (liver fluke) and ticks (a number of tick-borne diseases).

## Early Identification of Disease

A number of factors will contribute to whether or not a sick animal is identified:

- Skilful stockmanship.
- Regular and thorough observation of the group. This is easy where animals are housed, but may be difficult with extensively managed cattle at pasture. It may be worth while to offer a small amount of concentrate daily so that the whole group is seen each day.
- Close communication between stockpersons.
- Familiarity with the herd and the individuals within it. Whereas the regular herdsperson may easily pick out a sick cow, the relief herdsperson – even if very experienced – may not do so simply because he/she does not know the animals so well.
- Herd monitoring of disease status. This includes milk somatic cell counts (SCC) as a measure of sub-clinical mastitis, bulk milk testing for antibodies to diseases such as leptospirosis, infectious bovine rhinotracheitis (IBR) and bovine virus diarrhoea (BVD), and abattoir figures of liver rejections for fluke from beef cattle.
- Monitoring of nutrition and production. Disease may be caused by incorrect or inappropriate nutrition, but also the appetite of animals affected by disease is often reduced. Disease is also likely to affect production, for example growth rate, milk yield or fertility. Monitoring nutrition and production allows changes in feed intake or production level to be discovered, which may provide early warning of emerging disease.

## Investigation of Disease

Specific measures for controlling disease can be put in place only after a diagnosis has been made, and thorough examination of the sick animal generally provides the easiest method to reach a diagnosis. Diagnosis of disease at individual or herd level may include the following steps:

- Clinical examination, following the checklists detailed on pages 195, 196 and 199.
- Examination of herd and individual records (as above).
- Sampling. Specific diagnostic samples (for example, blood, milk or faeces) may help identify the cause of disease.

- Test treatment. This may be a cost-effective way to confirm a highly likely diagnosis, but it is often not appropriate if the likely diagnosis is not clear. For example, it is costly to treat a group of scouring calves at pasture with a wormer; if the cause is actually coccidiosis, calves may become much more sick before it is clear that the wormer has been ineffective. A more cost-effective approach would be to take dung samples from calves and treat according to laboratory findings.

## Treatment of Sick Animals

Disease treatments include non-specific and specific treatment measures.

### Non-Specific Treatment

The purpose of non-specific treatment is to relieve the clinical signs of disease so that the animal feels better and is better able to fight disease. This includes:

- Provision of palatable fresh food.
- Provision of water, with the appropriate addition of electrolytes to correct or prevent dehydration.
- Vitamins. The debilitated animal that is eating little may quickly become depleted of essential vitamins. Also, most vitamins are metabolized in the liver, and liver function is poor in many sick animals, particularly in the high-yielding dairy cow.
- Rumen stimulants. Rumen function is often poor in sick cattle and appetite will not return until it is re-established.
- A dry, draught-free bed and heat via a heat lamp for the young calf.
- Nursing care.
- Pain relief (if required).
- Antibiotic cover (where appropriate) to prevent secondary bacterial invasion of a debilitated animal.
- Isolation to allow the animal to rest without having to compete for feed, water and space.

*The veterinarian makes a full clinical examination and takes samples to investigate cause of disease.*

*Specific Treatment*

Specific treatment will, of course, depend on the diagnosis that has been made (*see* above). Success of treatment depends on making the right choices (with veterinary assistance as required) of drugs that likely to be effective (and with minimal side-effects), treating early rather than waiting until animals get very sick, and providing supportive care as above. The efficacy of medicines stored on farm can be maintained if you:

- Store medicines in a cool, dark place or re-frigerate if necessary.
- Give a full course of medicine at the correct dosage for animal's weight and age. Weigh animals if necessary.
- Use medicines before their use-by date.

Human safety can best be protected if you:

- Store medicines in a locked cabinet.*
- Dispose of needles in 'sharps' container, and send used bottles and syringes for disposal by your veterinarian.*
- Avoid accidental self-injection by making sure animals are well-restrained.
- Use protective clothing (gloves etc) as required.
- Adhere to the withdrawal times for meat and milk so that no residues enter the food chain.*
- Record use of medicines, together with batch numbers and identity of animals treated.*
(* Mandatory measures.)

*Antimicrobials / Antibiotics*

These are used to combat infection by bacteria or mycoplasmas. There is a multitude of antibiotics on the market, and it is not appropriate here to discuss the various attributes of each. The vet will advise on choices for particular diseases, but it is important to understand the factors involved in those choices.

1. They work in different ways, and each is particularly effective against certain

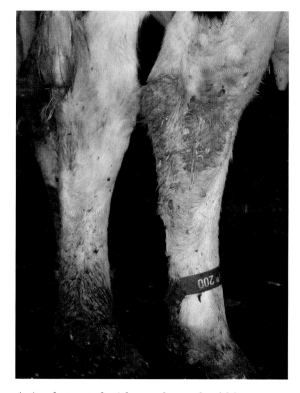

*Animals treated with any drugs should be clearly marked so that withdrawal times can be observed.*

classes of bacteria and may be completely ineffective against others.

2. They are variably distributed within the body: for example, one may better penetrate the udder and another the lung, so would be appropriate for use in mastitis or pneumonia, respectively.

3. Some can penetrate tissue cells to destroy bacteria.

4. The duration of action may be short or long. Long-acting preparations are appropriate if it is difficult or stressful to handle animals for repeat treatments.

5. Withdrawal times have been calculated to identify how long the antibiotic stays in the animal. A short withdrawal time is

useful when treating a high-yielding dairy cow or a beef animal that will soon be due for slaughter.

A sick animal may not improve despite antibiotic treatment. This may be because the bacteria have developed a specific resistance to the antibiotic (like 'superbugs' such as MRSA seen in human medicine). Incomplete courses of antibiotic or dosage inadequate for the weight of animal predispose to bacterial resistance to antibiotic because a low dose or incomplete course kills all the most sensitive bacteria, leaving behind those that are more resistant. In some cases, the antibiotic cannot penetrate the affected tissue – for example, where there is inflammatory fluid (in a lung with pneumonia) or scar tissue (in an udder chronically infected with mastitis). Antibiotic may also fail if an incomplete course of antibiotic is given, so that treatment is stopped before all the bacteria are gone and they multiply again to cause a relapse of clinical signs. In some cases an animal fails to respond when the treatment is not started until the animal is irreversibly damaged by disease.

## Specific Control Measures

The aim of disease control may be either to reduce the incidence of disease in the herd such that it no longer causes significant losses or to eradicate the disease completely.

### Control to Reduce Incidence of Disease

The principles discussed for prevention of disease (*see* page 184) also apply here, but disease-control measures include specific action to tackle specific problems on a unit. For example:

- Vaccination.
- Anthelmintics (drugs used in the control of parasitic worms).
- Dry cow therapy (DCT).
- Mineral and trace element supplements.
- Diets for the dry cow and transition cow (for three weeks before calving until after calving) to prevent metabolic disease.

VACCINATION

Once a disease has been identified, it is often cost-effective to vaccinate against the causative organism(s). In some cases (such as ringworm), clinical signs are enough to make a diagnosis, and vaccination can proceed. In other cases (such as calf pneumonia), a number of different pathogens may be causing disease. Vaccination will reduce incidence only if the vaccine protects against the pathogen involved. Further diagnostic testing (often by blood sampling) is needed to identify which causative organism(s) are involved. It is essential also to identify other risk factors. Using the example of calf pneumonia again, if environmental conditions are poor and calves of different ages are mixed, the challenge may be so high that even vaccinated calves will succumb.

ANTHELMINTICS

Calves meet infective worm larvae on the pasture. If numbers are low (with very low stocking density or new ley pasture), the worms cause little or no check in growth. However, on many units it is necessary to dose calves with anthelmintic during their first season at pasture. This can be done using a bolus that releases the anti-parasitic drug for a prolonged period or as a series of pulses, protecting the calf for the entire grazing season. Alternatively, calves can be dosed several times during the season, or dung samples can be collected and calves dosed only if the worm-egg counts rise. The advantage of using worm-egg counts is that calves are dosed only when they need it; this avoids the expense of using unnecessary drugs and, by reducing the number of times that calves are treated, may help to delay the onset of resistance to anthelmintics. However, relying on worm-egg counts can be dangerous where the risk is high – disease may occur before worm eggs are found in the dung.

DRY COW THERAPY (DCT)

This is antibiotic formulated for use within the udder (intra-mammary) and helps to reduce the number of cases of mastitis in three ways:

Larvae develop into adult
worms in calf/adult gut

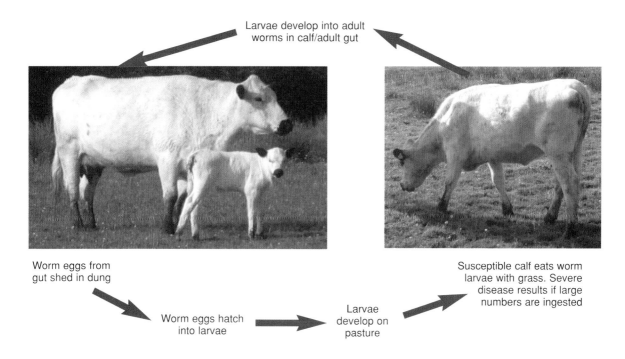

Worm eggs from
gut shed in dung

Worm eggs hatch
into larvae

Larvae
develop on
pasture

Susceptible calf eats worm
larvae with grass. Severe
disease results if large
numbers are ingested

*Schematic diagram of the life cycle of gut worms in cattle.*

Larvae migrate from gut to lungs
where they develop into adults
and produce larvae

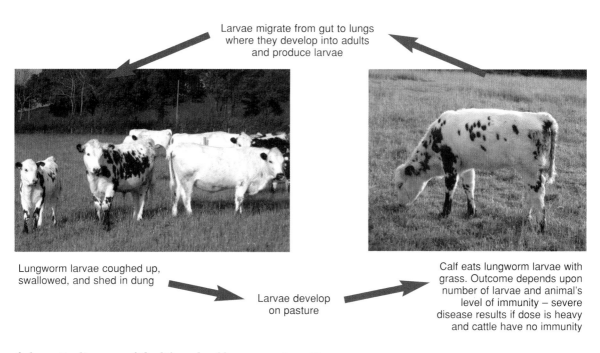

Lungworm larvae coughed up,
swallowed, and shed in dung

Larvae develop
on pasture

Calf eats lungworm larvae with
grass. Outcome depends upon
number of larvae and animal's
level of immunity – severe
disease results if dose is heavy
and cattle have no immunity

*Schematic diagram of the life cycle of lungworm in cattle.*

1. It is much more effective at eliminating mastitis-causing bacteria than treatment of chronically infected cows during lactation, thereby reducing the pool of chronically infected cows in the herd.

2. It reduces the number of new infections that occur during the dry period (often causing mastitis in early lactation).

3. It aids in the prevention of so-called 'summer mastitis', which occurs in dry cows, often in summer.

MINERAL AND TRACE ELEMENT SUPPLEMENTS

Cattle can be dosed directly with slow-release supplements that provide the animal with its requirements for weeks or months. This can be done by mouth, with a solid bolus that lodges in the rumen or by a long-acting (depot) injection. Magnesium and selenium can be offered in water troughs, but this is suitable only where cattle have no access to alternative water sources (such as streams), which they may choose instead. Minerals and trace elements can also be provided in feed or as a mineral lick. A mineral lick has the advantage of being easy to use, but may not be as effective as alternative methods because not all cattle take the same quantity. This is particularly important in the case of magnesium, where cattle require a daily intake.

DIETS FOR THE DRY COW AND TRANSITION COW

These are given for three weeks before calving until after calving to prevent metabolic disease. Improved understanding of the metabolism of the dairy cow has led to development of rations for the cow, designed to prepare her for the next lactation. These can be very effective in control of milk fever, fatty liver and ketosis. (*See also* Chapter 5.)

## Eradication of Disease

Eradication of infectious disease may be carried out in the individual herd or at national or international level. At first glance it appears an attractive option. However, it is im-

possible to eradicate any disease caused by organisms commonly present in the environment or within the animal's body, or where disease is multi-factorial, such as calf pneumonia. UK history shows that eradication of disease such as foot and mouth disease (FMD) and bovine tuberculosis may be both costly and difficult. In general, eradication of disease is based on identification of infected animals and their removal from the herd. If it is possible to distinguish vaccinated cattle from infected ones (for example, BVD and IBR), the number of infected cattle can be reduced by vaccination in conjunction with eradication.

## Biosecurity

This defines the measures that can be taken to protect the herd from threats posed by incoming disease. The effect of disease introduction depends on the disease itself and the immunity of the herd to that disease. The worst-case scenarios are when disease spreads easily, causing large financial losses, and when there is no immunity within the herd prior to its entry. A classic example is foot and mouth disease, but there are many common production diseases that can have serious economic impacts. Despite this, awareness of the importance of good biosecurity practices is poor on many farms.

Disease may be introduced through other cattle, humans (via clothing, boots, skin, and so on), vehicles, wildlife, feed, and even semen and embryos.

### Purchased Stock

Most disease entering a herd results from close proximity to, or introduction of, infected stock. Ideally the herd is 'closed', i.e., there is no purchase or hire of any stock, all replacements being reared on farm. This is the best way to keep out disease, but in practice there are difficulties, for example:

- Inability to avoid contact with neighbouring stock.
- A need to expand herd size above the ability to produce home-bred replacements.

*Double fencing provides an effective barrier between neighbouring cattle. The two fences should be 3m apart to prevent contact. In this example the fences also protect a newly laid hedge.*

- Higher than anticipated culling rates, resulting in the requirement to buy replacements to maintain herd size.
- The need to use a bull as an aid to fertility management.
- The requirement to breed replacements reduces the opportunity to bring in new genetic material and ties up resources that could otherwise be used for production.

.
Therefore most herds need to introduce some stock.

Not only may disease be introduced by incoming stock but the existing herd may transmit disease to purchased animals. To prevent this, it is important to know the disease status of both the farm's own herd and the vendor's herd. The disease status of the herds can be discovered through routine sampling and/or herd vaccination or by specific diagnostic testing prior to purchase. This should be discussed with the veterinary surgeon, as not all tests are entirely reliable. Knowledge of the geographical distribution of disease can also be helpful. For instance, in the UK certain parts of the country are known to have higher numbers of herds infected with bovine tuberculosis than others, so avoiding purchase of stock from such areas is a sensible precaution.

Purchased cattle should be isolated on arrival to allow further tests and/or treatments to be conducted prior to meeting the herd. In dairy herds this means avoiding purchase of lactating cows, as it is usually impractical to isolate cows that need to be milked.

### Perimeter Fencing

Perimeter fencing must be strong, well maintained and stock-proof, with double fencing at least 3m apart to avoid direct contact from other cattle. Electric fencing can be used as a temporary measure. An alternative is to use perimeter fields adjacent to other stock exclusively for arable or cropping purposes.

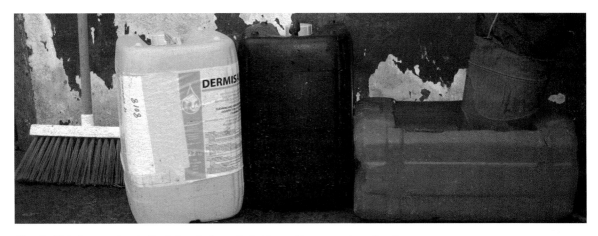

*Fresh disinfectant boot dip helps to keep infectious disease out and raises awareness among staff and visitors of the importance of biosecurity.*

### Water Courses

Water, particularly slow-moving water that runs from farm to farm, can act as a vector of infectious disease. Stock should be prevented from drinking such water.

### Spreading of Slurry

Diseases such as salmonellosis or Johne's can potentially be spread by infected slurry on grazing land. Ideally, slurry should be used on land that is destined for production of arable crops or conserved forage, or it should be stored for eighteen months or two years to minimize the risk (infectious organisms die during storage).

### Other Farm Species and Wildlife.

Not all diseases are specific to one host species, and there are several examples of important production diseases that can cross species barriers. Risks can be reduced by measures such as preventing entry of wildlife into stock buildings, use of vermin-proof feed stores, raised water troughs and fencing off badger latrines.

### Vehicles and Personnel

Vehicles that travel from one farm to another may transfer disease on dirty wheels etc, and personnel, whether staff or visitors, may carry organisms on clothes, boots, hands, and equipment. Ideally, vehicles are kept outside the farm and certainly away from the herd. A boot dip with fresh disinfectant and a brush should be provided. Overalls can also be provided to reduce the risk of brought-in disease.

## KEY SIGNS OF HEALTH AND DISEASE

It is essential for any stock-keeper to be able to recognize the health status of cattle. Here we focus mainly on the normal, healthy animal because:

1. There are many more normal than abnormal cattle, so it is much easier to become familiar with what is normal, and

2. There is, broadly, only one set of normal parameters, while there is an almost infinite range of signs that indicate abnormality or disease, and their interpretation can be complex. For example, a group of animals may be dirty because they are housed in poor or overcrowded conditions, or because they have chronic diarrhoea, or because they are debilitated and do not groom themselves.

However, there is also a range of normality, and it is useful to get to know individual animals so that any change from normal (in behaviour, milk yield, appetite, demeanour, and so on) can be recognized early. Equally, there is a gradation between healthy and sick: mild clinical signs or changes early in the progress of a disease may be very subtle, while signs such as collapse, frothing at the mouth, convulsions or death will be noticed by even the most unobservant stockperson.

When looking for the signs of health and disease, it is essential to be thorough, and we have used a systematic checklist approach so that nothing is missed. This will help prevent the possibility of jumping to conclusions that may be wrong. For example, it is easy to assume that a freshly calved cow that cannot get up is suffering from milk fever, but in fact she may have a toxic mastitis or infection of the uterus, and delay in providing the correct treatment may be fatal.

## Signs of Health

There are five steps for making a thorough investigation of the signs of health or disease that an individual animal or group displays:

1. Clinical history of the animal(s).

2. Observation of the group from a distance, either during daily interactions with the herd or as part of a specific investigation.

3. Approach towards the group to observe individuals more closely.

4. Estimation of feed intake and production rate (growth, milk yield, etc.) either by direct observation or from recorded data.

5. Detailed examination of one or more individuals.

---

### Group Observation

By observing cattle from a distance, the experienced stockperson will be able glean important information that will help to indicate the herd's state of health. The factors to consider and signs to look for include:

**General behaviour**
- Cattle mostly act together, feeding at the same time, cudding at the same time.
- Small herds at grass tend to face in the same direction and move together around their range (not possible with large herds on restricted grazing).
- Use of time (varies from calf to adult):
  Feeding.
  Cudding.
  Sleeping.
  Idling.
  Other.
- Range of normal social behaviour, such as aggression or fighting, vocalization, playing.
- Nature of cudding: vigour of chewing; number of chews per cud; cud should not be dropped.
- Behaviour in the milking parlour, such as: order of milking (may come in late instead of early); restlessness in parlour; may not eat.
- When food is offered, should approach immediately to feed.

**Locomotion**
- Free movement of all four legs with even gait.
- Flat (not arched or hunched) back.
- Stand square on all four feet.
- Rhythmic movement of the head with each stride.
- No reluctance to move.

**Lying down**
- All but young calves are usually cudding when lying down.
- Cattle often sleep with head curved round against the flank. (Adults spend little time sleeping; young calves sleep for several hours per day.)
- Forelegs are generally tucked underneath.
- Cattle seldom lie flat out on their sides.
- They should lie down and rise without difficulty.

**Demeanour**
- Alert and responsive to movement or noise (follow sound and movement by turning head).
- Ears up (not drooping).
- Head up (not drooping).

## Close Examination

Having studied the animals from a distance, the stockperson should approach the herd to make a closer examination:

**Eyes**
- Bright, moist, clear (transparent), with no discharge.
- Neither sunken nor bulging.

**Facial expression**
- Symmetrical.
- Ears up.

**Nose**
- Moist/wet, with clear mucus.
- No discharge (blood or pus).

**Coat**
- Glossy.
- Smooth (according to season and breed).
- Dense/thick.
- Even.
- Clean (within limits imposed by bedding). There should be no signs of recent or chronic diarrhoea or dirty vaginal discharge.
- No crusty or raw skin lesions.

**Body condition score (BCS)**
- Should be appropriate for stage of production.
- Range of BCS should be small (ideally not more than one score) within group.

**Growth of calves and youngstock**
- Should be fairly even.
- No stunted animals.

**Hooves**
- Even, not overgrown.

**Rumen fill**
- Observe fill on left side, behind ribs. Should be neither swollen (bloat) nor empty.
- Calves should not be pot-bellied.

**Respiration**
- Quiet, slow, even, and almost silent.
- Respiratory rate increases with heat stress or fear.
- No coughing.

**Belch gas**
- Regular, about once every one to two minutes in ruminating cattle.

**Dung**
- Range of normality according to feed ration and stage of production consistent among group.
- Ideally a good 'pat' – not too firm, not too soft.
- Little or no undigested material.
- No blood or mucus.
- Normal quantity.
- Dung should be passed without straining or signs of pain.

**Urine**
- Clear; pale to straw-coloured.
- No blood.
- Urine should be passed without straining or signs of pain.

*Clinical History of the Animal(s)*

This includes all information that may potentially affect health, for example:

- Age.
- Weaned/not weaned (calves).
- Pregnant/not pregnant and stage of pregnancy.
- Recent service (AI or bull).
- Calving date.
- Lactating/non-lactating.
- Home bred or bought in.
- Recent (within the last thirty days) cattle purchases to the unit.
- Changes in management or nutrition within the last ten days.
- Housing.

*Observation from a Distance*

Observing a group of cattle from a distance enables the stockperson to assess the animals' behaviour when they are undisturbed and uninhibited by human presence. This can be invaluable in monitoring health.

*Observation on Approach*

The way the animals respond to human approach is also an indicator of their state of health.

head up, alert and curious, with bright eyes and pricked ears

no discharge from eyes

nose moist; no discharge

stands square; no lameness

good body condition

no signs of diarrhoea

rumen full, not bloated or empty

coat thick and glossy

*The healthy animal.*

*This cow is sick. Her ears droop, her head is down, she is not interested in her surroundings, her posture is odd, with hind legs stretched back, and she is dehydrated.*

*Nasal discharge in a calf is usually caused by viral infection.*

The normal animal looks up when a human approaches, generally turns towards the human, and follows the approach with ears and eyes. It moves away as the human steps inside the flight distance (*see* Chapter 8, page 176). Flight distance varies enormously according to the herd's experience of humans: in the case of quiet dairy cows, or calves that are bucket-fed, the flight distance may be nil; in extensive beef animals who have seldom been handled it may be 40-plus metres. However, there is a normal flight distance for any group, and it usually indicates that there is a problem if a nervous animal allows a human to approach or if a quiet one charges away or attacks.

On approach, it is possible to make closer physical observations of the animals:

### Estimation of Feed Intake and Production Rate
Estimations of growth, milk yield, and so on, can be made either by direct observation or from recorded data. It is often a change in milk yield or quality in the dairy herd, or a fall-off in growth rate of heifers or beef animals, that alerts the stockperson to an emerging problem, either in an individual animal or in a group. Clearly, the better the available records, the more accurately such problems can be pin-pointed. Fall in level of production does not always indicate emerging disease; it may reflect a nutrition-related problem or even an environment-related one such as heat stress.

Equally, an individual cow may stop finishing her concentrate in the parlour, or a group may no longer clear their daily ration of cereal. Such changes in feed intake do not always indicate disease: they may be caused by heat stress or they may be related to the animal's stage of production, or they may follow a change in ration palatability. A fall in appetite usually indicates a problem and should be investigated.

### Detailed Examination
If everything appears normal so far, there is generally no need to make a more detailed examination. However, if a problem is suspected, it is essential to look more closely at

## Detailed Examination of the Individual Animal

### Temperature

The thermometer should be shaken down (mercury) or turned on (digital). It should be inserted into the rectum and pressed lightly against the rectal wall. It is important to make sure that the thermometer is not embedded in dung or in a pocket of air, which will give an erroneous reading. Normal temperature in the calf is 39°C (101.5–102.5°F); in the cow it is 38.5°C (101–101.5°F).

### Rumen contractions

If a closed fist is pressed over the rumen of the normal animal, a rolling contraction should be felt regularly one to two times per minute.

### Navel

The calf's navel should be dry within twenty-four hours of birth, shrivel within seven to ten days, and disappear shortly after that. The navel should not be swollen or painful, and there should be no pus.

### Udder and milk

The udder should be soft, even and pain-free. Milk should be normal, white and without clots or smell.

Colostrum, produced for up to four days after calving, may be very thick and creamy yellow in colour. It should not contain clots, and the udder should not be painful.

There may be some swelling in front of the udder before calving.

### Foot

If a foot is to be lifted, this should always be done with the animal in a crush. Special foot-trimming crushes give excellent access to both front and rear feet, but otherwise care must be taken to avoid injury to animal or operator.

Unless the animal is extremely quiet a belly-band (a wide strap placed under the animal's belly) should be used to prevent the animal from going down. The belly-band should not be tightened so much that it suspends the animal, as this can damage her. The leg is lifted with a strap fitted above the hock, ideally using a winch, and taking care not to stand directly behind the animal. It may be necessary to prevent the animal from kicking by strapping the leg to the side of the crush. A quick-release knot should be used so that the animal can be freed quickly if it struggles (otherwise it may damage itself). Once raised and immobilized, the foot can be examined. Common problems include:

- Ulcerated lesions between the claws.
- Raw lesions at the junction between claw and skin (the coronary band).
- Overgrown hooves.
- Solar ulcers: part of the sole is eroded in a circular, often infected, ulcer.
- Lesions at the junction between the sole and the wall of the hoof (the white line).

Foot-trimming and lameness treatment are skills that require a lot of practical training, and in most cases expert help will be required to deal with foot disease.

### Swellings and lumps

There should be none.

---

individual animals. It may be obvious that a particular animal should be singled out, otherwise a representative range (across the spectrum of, for example, age, stage of pregnancy or lactation) should be selected.

Detailed examination requires adequate restraint both for the safety of the operator and the animal. Young calves can be immobilized in the corner of a pen – once they realize that escape is impossible they quickly stop struggling. Older animals need to be restrained in a crush. Ideally this should lock the head so that there is no need for a bar behind the animal, and should open to enable safe examination of the various parts of the animal (udder, legs, head). A halter is needed for detailed examination of the head (*see* Chapter 3), or a pair of 'bull-dogs' can be used. These clip across the sensitive nasal septum so that the animal, in order to avoid pain, does not move its head.

Use all your senses to make your examination:

- Vision: to look closely at every part of the animal.

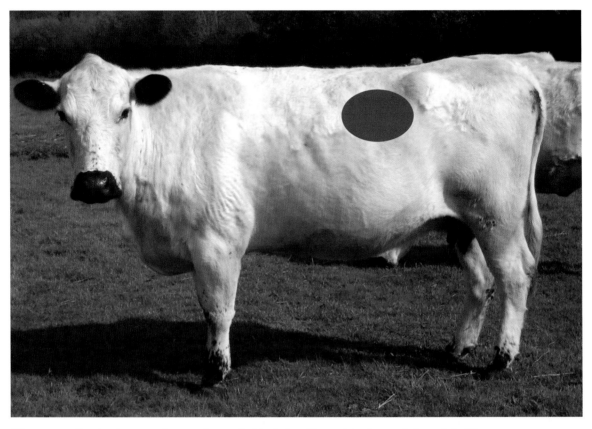

*The rumen lies in the area shown above – behind the ribs and in front of the pelvis. You can observe rumen fill and whether there is bloat. If you push a fist over the rumen you can feel its contractions (about one per minute).*

---

### Veterinary Examination

In addition to the detailed, individual examination outlined in the box on page 199, a veterinary surgeon will conduct a clinical examination, commonly using the following techniques:

**Stethoscope**
The stethoscope amplifies internal sounds, and is used to examine:

- Rumen movement.
- Gut movement and contractions.
- Build-up of gas in abdominal organs – for example in the displaced abomasum (LDA or 'twisted stomach').
- Heart sounds.
- Lung sounds.

**Rectal examination**
This allows the vet to explore the organs in the back part of the abdomen (belly), particularly:

- The reproductive tract (for example for pregnancy diagnosis).
- Some parts of the gut (much is out of reach).

**Vaginal examination**
This enables the vet to examine the cow for signs of internal infection or to assess progress during calving.

**Ultrasound**
This is used in conjunction with the rectal examination to provide more detailed information about the reproductive tract.

*A young calf that becomes sick must be identified swiftly or it may die. This calf's ears are drooping, her eyes are dull, and her posture is hunched, which is often an indicator of abdominal pain.*

- Hearing: to listen for normal breathing (should be very quiet).
- Smell: to recognize the unpleasant odour of (for example) a discharging infection from the uterus or mastitis milk, or to check for the presence of ketones on the breath.
- Touch: to explore the consistency of the udder and teats, to feel for contractions of the rumen and to explore any abnormal swellings or lumps.

The veterinary surgeon will make further examinations of the individual animal.

## Signs of Disease

In broad terms, any deviation from the set of normal parameters outlined above may indicate a problem. This might be anything from a vague set of signs that suggest an animal is not in optimal health, to severe clinical signs characteristic of a particular disease.

There are far too many disease signs for them all to be itemized here, but common ones include:

- Diarrhoea. There is increased production of loose dung with or without blood and mucus. The tail and hindquarters may be dirty, and the animal may show signs of abdominal pain and/or strain to pass dung.
- Lameness. The animal limps on one or more legs, may be reluctant to move, and its back is usually arched. Lameness is common in dairy cows, especially in some herds, and the number of affected animals is often underestimated by farmers. Level of herd lameness can be assessed using a 'locomotion score' to identify animals whose gait is abnormal.
- Coughing. This is usually associated with increased respiratory rate and noise. It usually gets worse when the animal moves or runs. There may be discharge from the eyes and nose.
- Fever. The temperature is raised, respiratory rate is usually increased, appetite is reduced, and the animal may sweat. Animals with fever are usually dull and often separate from others in the group.

*Diarrhoea is common and its cause must be investigated by laboratory examination so the correct treatment and preventive measures can be chosen.*

- Nervous signs. There are many different signs including profound dullness, blindness, muscle twitching, incoordination, collapse, and bellowing.
- Bloat. The belly is distended, more commonly on the left. Acute bloat is very serious and the affected animal generally requires immediate veterinary attention.
- Mastitis. The milk changes, becoming watery or clotted, and the affected quarter may be swollen, hard and painful.
- Dehydration (*see* box).

## CONCLUSION

The principles of disease prevention and control are the same in any herd – they are based upon sound husbandry and stockmanship. The details, however, vary enormously from one herd to another. Each must therefore tailor a set of prevention and control policies – such as vaccination programmes, introduction of hygiene measures, care for the newborn calf, and biosecurity – that are specific to the individual farm and take account of factors such as herd size, location, production level, and type of stock.

### Dehydration

There is a gradation of clinical signs according to the severity of dehydration, including: depression (the animal appears dull and listless), ears down, dry nose, and reduced appetite. More specifically there is skin 'tenting'. In the normal animal the skin is elastic and flexible so that if it is pinched and released, it immediately snaps back flat. If the skin of the dehydrated animal is pinched, it remains tented up and only slowly returns flat. The eyes are sunken – this is more obvious in animals that are severely dehydrated.

Dehydration commonly occurs as a result of diarrhoea or toxic mastitis, or after a difficult or prolonged calving, but any animal that is sick is likely to become dehydrated, particularly if its temperature is raised. Dehydration, even if mild, causes the animal to lose its appetite and become depressed, and severe dehydration may lead to death. It is therefore important to recognize the signs early and to prevent the problem wherever possible in order to promote recovery.

In order to counteract dehydration, the animal should be offered as much oral electrolyte solution as it will drink (ideally this should be warmed). An adult animal may require up to 60 litres of water or electrolyte solution. If unwilling to drink, a stomach tube should be used to provide fluids: 2–5 litres for calves and 15–30 litres for adults. Intravenous fluid therapy via a drip may also be used for valuable animals, and is a cost-effective treatment for the calf with diarrhoea.

# CHAPTER 10

# Cattle Welfare
# and the Law

Animal welfare has been mentioned frequently throughout this book, and concern for animal welfare is implicit in many of the recommendations about feeding, housing, disease control, and so on. It could therefore be argued that a chapter devoted to the subject in isolation is superfluous, but it has been included because animal welfare is an issue of increasing importance in livestock farming, and because the science that underpins our understanding of animal welfare can inform the way we keep cattle.

There is plenty of scientific evidence to confirm that mammals are sentient beings

*Cattle that have been handled show little fear of humans and plenty of curiosity. This is apparent in these animals, which are keen to investigate us with our camera. (Note that the loose wire is a significant safety risk.)*

203

with the capacity to suffer. Anatomical and physiological studies have shown that their nerve pathways are very similar to those of humans – for example the flinch an animal makes as a needle goes into the muscle to deliver an injection (pain response). Behavioural investigations have demonstrated that animals make choices and will work to achieve rewards such as food or a comfortable bed or a reproductive partner. They also learn from experience. Anyone working with cattle can confirm these scientific results every day – the bulling heifer who breaks out of her field to find a bull (seeking a reproductive partner) or cattle charging towards an ATV (to achieve reward, having learnt from experience that it is carrying their concentrate feed). We cannot, of course, conclude that animals feel things in the same way that humans do, only that they try to avoid suffering and make choices to improve their well-being. It is therefore morally right that we should try to minimize suffering and promote well-being in livestock.

Even if the moral argument for animal welfare were not convincing, there is also the carrot of premium prices for 'welfare-friendly' produce and the stick of welfare legislation that protects farm animals. Both are the result of society's concern about animal welfare, which has increased progressively in the world over the last forty years. This concern (together with worries about food safety) has led to the development of niche markets for livestock products from animals kept on less intensive units, and to the setting up of Farm Assurance Schemes (FAS). FAS have been developed by supermarkets and other food marketing organizations to offer the public reassurance about the conditions under which farm animals have been reared and managed. Farmers who fail to meet the standards demanded by the FAS may lose their market. The law, which at its best reflects society's values, regulates the ways in which livestock must be treated on the farm, at market, during transport and at slaughter, and penalizes those who fail to comply.

The first part of this chapter explores the welfare effects of various aspects of cattle husbandry as it is commonly practised on today's livestock units. A section at the end of the chapter examines the legal protection of cattle welfare in the UK, together with some other laws important to cattle keepers.

## THE FIVE FREEDOMS

The Farm Animal Welfare Council (FAWC) has published a framework of ideal welfare standards known as the Five Freedoms. These are:

1. Freedom from thirst, hunger and malnutrition by ready access to fresh water and a diet to maintain full health and vigour.

2. Freedom from discomfort by providing a suitable environment including shelter and a comfortable resting area.

3. Freedom from pain, injury and disease by prevention or rapid diagnosis and treatment.

4. Freedom to express normal behaviour by providing sufficient space, proper facilities and company of the animal's own kind.

5. Freedom from fear and distress by ensuring conditions that avoid mental suffering.

Clearly all of the freedoms will be compromised on poorly managed farms or where cattle are neglected or cruelly treated. However, in the process of commercial cattle farming there are factors that tend to promote or limit each of the Five Freedoms even on the best-run farms, and these are explored below.

### Freedom from Thirst, Hunger and Malnutrition

Good stockmanship is essential to ensure that plenty of appropriate, good-quality feed and fresh water are always available. This includes pasture management and forage conservation as well as ration formulation and monitoring of animal condition and performance. Sick animals recover more quickly

*Welfare of livestock is a subject of increasing concern to the general public, and the farmer who manages cattle well, under extensive conditions like these, can command premium prices for the beef he sells.*

if supplied with palatable feed and fresh water, so sick pens should provide these. Cattle performance is optimal if they are well fed and watered, so there is a strong eco-nomic benefit in providing good nutrition and water.

However, since it is cost effective to have spring-calving beef cows gain weight during

*Freedom from hunger. Plenty of good-quality feed, appropriate to the class of stock, should always be available.*

*Freedom from thirst. Cattle need access to fresh water at all times.*

the summer at pasture, and then lose weight over the winter months (when forage is expensive), it is likely they will be hungry. Out-wintered cattle may also be hungry and short of water in bad weather. The energy demands for peak lactation are so high for high-yielding dairy cows that they cannot physically consume enough to meet them. An energy deficit is inevitable, and this is known as 'metabolic hunger'. Mild, sub-clinical deficiencies in minerals and trace elements are very common.

## Freedom from Discomfort

Prolonged discomfort leads to stress-related disease, so every effort should be made to avoid discomfort. Good stockmanship is required to keep animals comfortable, for ex-

*Freedom from discomfort. Cattle should be able to choose shade in summer and shelter in a storm.*

*Freedom from pain, injury and disease. This cow is very sick and looks completely miserable.*

ample by provision of adequate bedding and frequent mucking out.

Young calves may feel cold because they are not able to regulate body temperature as well as older cattle, while heat stress is common in high-yielding dairy cows because of their high metabolic rate. Poor cubicle design causes discomfort and leads to hock lesions and lameness, but the capital cost of replacing them is high. Cattle transport usually involves some discomfort but, while every effort should be made to minimize this (for example by slow, careful driving), transport – and its associated problems – cannot be avoided altogether.

## Freedom from Pain, Injury and Disease

The cost of injury and disease is high so should be minimized. Well-managed farms have action plans in place to prevent disease (such as vaccination). Good stockmanship is required to identify disease early and take appropriate action. Early veterinary treatment minimizes the effects of pain, injury and disease, and euthanasia of hopeless cases prevents further suffering (humane euthanasia is not a welfare issue). Local anaesthesia for routine operations such as castration minimizes pain.

However, the incidence of lameness, mastitis and pneumonia is high on many units because these are multi-factorial problems that are difficult and/or costly to control. There is, for example, little routine use of pain relief post calving or after dehorning.

## Freedom to Express Normal Behaviour

Small suckler beef herds at pasture have the optimal opportunity to express their full range of normal behaviours. For example, calves remain with their mothers for several months, and they can choose their range of activities and where to spend their time. They tend to move together in a group, grazing, cudding or resting at the same time. Growing calves (which spend less time feeding) also play.

Housing restricts normal behaviour, especially if groups are overstocked. Cattle are unable to express normal behaviour if they are kept in large groups with no established hierarchy, and smaller individuals within the group may be subject to bullying. High-yielding cows have no 'spare' time to rest; their metabolic demands mean that they must always be feeding, cudding or waiting to be milked.

*Freedom to express normal behaviour. A small beef herd at grass shows the full range of normal behaviour.*

## Freedom from Fear and Distress

Fear and distress may lead to stress-associated disease so should be kept as low as possible on every unit. Good handling facilities and quiet handling by knowledgeable stockpersons minimizes fear of unknown situations. A first-lactation group for calved heifers reduces stress and bullying in the dairy herd. The use of sedation for major veterinary interventions eliminates associated stress. Welfare training of abattoir and market staff has reduced fear and distress and improved meat quality.

Large group size leads to bullying, but is an integral part of management in large herds. Cattle that have been extensively reared are likely to be highly distressed when they are

*Freedom from fear and distress. First-lactation heifers kept in a group separate from cows are not bullied by older, dominant cows.*

*This farmer shows empathy with his cow, who is unwell.*

handled. Transport, markets, and abattoirs all potentially cause fear and distress, but this can be minimized by good stockmanship and handling. Early weaning of calves inevitably distresses both cows and calves.

Two factors are strongly linked to every aspect of cattle welfare:

1. Quality of stockmanship.
2. Economics of cattle production.

## QUALITY OF STOCKMANSHIP

Quality of stockmanship is probably the single most important determinant of cattle welfare, and embraces a range of skills including:

- Technical skills. Technical skills are essential, for example to ensure that cattle are fed and managed correctly at each stage of the production cycle and to identify and treat sick animals. These skills may be gained at home on the farm, but may also be acquired through training courses and from discussion groups and publications.
- Observational skills. Good stockpersons are observant and pay attention to detail, so they recognize the very first signs of disease, or subtle changes in quality of forage, or hazards that might cause injury.

- Empathy with stock. This stems from the rapport that a stockperson has with his/her animals. Those with good technical and observational skills are able to manage cattle well, but those who also have high empathy with stock will take extra care to make sure that animals are comfortable and to protect their welfare.

These skills are interlinked. The example of the calving cow illustrates this. The stockperson's *technical* skill is required to make sure that cows are adequately prepared for calving – at the correct body condition score, being fed an appropriate ration to minimize the likelihood of metabolic disease, and housed in an environment that will facilitate healthy calving. *Observational* skills are used to identify the first signs of calving and to recognize any change from normal that might indicate a problem. Technical skill is once again needed to deal with such problems – when to wait and when to intervene – and specialist knowledge is essential if the stockperson is successfully to help the cow to calve without injuring her. *Empathy* is required to make sure that she is stressed as little as possible, to provide aftercare for cow and calf that will maximize recovery, and to nurse any animal that is down or sick after calving.

The quality of stockmanship is limited in many herds by lack of available time to provide individual animals with the care they need if their welfare is to be protected. This problem is likely to become more common as average herd size increases and profit margins are squeezed. In the UK there is often only one skilled stockperson in charge of 200-plus animals – other tasks are carried out by workers who have no experience or even interest in livestock. This is certain to compromise welfare.

## ECONOMICS

Economic factors will inevitably compromise some aspects of cattle welfare on even the best-run units. Overall profitability of cattle farming determines the level of capital expenditure for housing or milking parlour, as well as investment in health care, quality staff and day-to-day consumables such as bedding and feed. When profits are poor the farmer may be unable to pay for a new cubicle shed but instead rely on outdated ones that are too small for the cows, leading to lameness, injuries and lesions to hocks. Reduced use of straw for bedding makes a small financial saving but may have a major impact on cattle comfort. If the value of stock is low (as in male dairy calves in the UK in 1996–2006) the farmer may feel unable to justify the time to care for them or the cost to treat them if they are sick.

In general, incremental improvements in animal welfare can only be achieved at a cost. The cost of changing production systems (for example, by banning battery hen-houses for egg production) is high for individual farmers, but the extra cost to the consumer (for example for a box of free-range eggs) is low. It can therefore be argued that consumers should pay those extra costs if they want welfare-friendly livestock products. Although public concern about welfare is increasing, and it is likely that this market will develop, with the exception of free-range eggs there is currently only a limited niche market for most high-value, 'welfare-friendly' produce. Most people do not prioritize

animal welfare as a factor when choosing food, so improvements in cattle welfare must be achieved without significant increases in cost.

Fortunately there are many examples where cattle welfare and economics work in the same direction:

- Effective disease control and prevention.
- Provision of excellent nutrition.
- Quiet handling of stock for transport and slaughter.

However, there are also several fundamental tensions between cattle welfare and economics, for example:

- Early weaning of dairy calves.
- Large groups and high stocking density.
- One stockperson in charge of very large numbers of animals to minimize labour costs.
- Export of young calves.
- 'Cheap' treatment options.

These tensions between animal welfare and economics may only operate within limits. For example, the farmer who chooses cheap treatment or does not vaccinate does not save money if animals die because they were not given effective medication or were not protected by vaccination; high stocking rates do not make cost-effective use of available housing if overcrowding leads to diseases such as pneumonia or environmental mastitis; cheap sources of feed are useful only if production and health can be adequately maintained.

## MEASURING CATTLE WELFARE

If we accept that we should try to minimize animal suffering, we need to identify the factors that are important to the cattle as determinants of suffering and well-being.

Scientists have developed experimental methods to measure welfare objectively, recording the animal's physiological and/or behavioural responses to particular stimuli under controlled conditions. These methods are costly and slow, and each generally answers only one small question about animal welfare. Interpretation of results may be complex, but the techniques have advanced our understanding of the animal's responses to its environment.

It is also possible to measure cattle welfare on the farm. These methods are more practical, and rely upon our understanding of the conditions under which cattle thrive and survive:

- Inputs. The advantage of using inputs as a measure of animal welfare is that they are objective and easy to quantify and compare. Their disadvantage is that they give no indication of the cattle's response to their environment.
- Outputs. Outputs provide a more direct indication of cattle welfare. The prevalence of lameness and hock lesions or the number of severely dirty cows, for example, gives a much more reliable measure of the adequacy of housing conditions and the animals' freedom from discomfort, pain and disease than measurement of cubicle dimensions, straw use or stocking density.
- Records. Records of disease and mortality show a picture of cattle health and welfare over time, provided they are accurate.

Some argue that production data can be used to assess cattle welfare, on the assumption that only healthy animals with good welfare will grow, produce milk and breed. It is certainly true that cattle will produce optimally only if disease levels are low and if standards of nutrition, housing and management are good. However, it does not follow that lower production levels indicate poorer welfare. For example some farmers grow beef animals slowly or to produce lower milk yields because they have opted for more extensive or lower-input, lower-output systems. Some believe that very high milk production may, in itself, indicate a welfare compromise. Cow longevity may provide a better indicator of cattle welfare than other production data be-

---

**Measuring Cattle Welfare**

**Inputs**
- Quantity of feed available.
- Feed space per head.
- Water trough volume per head.
- Cubicle dimensions.
- Stocking density.

**Outputs**
- Condition score.
- Prevalence of disease (e.g. lameness, pneumonia).
- Dirtiness of coat.
- Behaviour towards humans (e.g. flight distance, *see* page 176).

**Records**
- Disease incidence.
- Mortality rate.
- Milk yield.
- Growth rate.
- Milk protein and fat level.
- Longevity/culling rate.

---

cause, provided only forced culls and deaths are included, it shows how well the cattle cope with their environment.

# INDICATORS OF GOOD WELFARE

Animal welfare is usually considered only in terms of suffering, the implication being that in the absence of signs of suffering animal welfare must be adequate. However, signs of cattle welfare do not need to be so negative: cattle also show positive signs of well-being. The animal welfare scientist might dismiss most of the following behavioiural signs as anthropomorphic, but at the very least they indicate that the animals are healthy and relaxed:

- An animal that stretches its back and hind legs after sleeping or lying down to chew the cud.
- Cattle, usually calves, that play together, often chasing one another and kicking up their legs.

211

*A heifer bounds about playing in fresh straw, showing a sense of well-being.*

- Cattle that play with new straw bedding.
- Cattle that scratch themselves on scratching brushes.
- Tame cattle that seek human contact.

## WELFARE LAWS

In this section the principal laws that currently apply to cattle farming in the UK are outlined, as well as the legal framework in which they operate. The laws are designed:

1. To protect animal health and welfare.
2. To protect human health.

For nearly a hundred years the most important animal welfare law in the UK was the Protection of Animals Act. This law made it illegal to cause 'unnecessary suffering' to any domestic animal, either by doing something or by omitting to do it. In 2006 it was replaced by the Animal Welfare Act (England and Wales), which updates and simplifies the law. It strengthens penalties and eliminates some loopholes. Importantly, it should reduce animal suffering by enabling preventive action to be taken before suffering occurs (where previously no preventive action could be taken).

The rather vague offence of 'unnecessary suffering' protects animals from neglect and deliberate acts of cruelty. More relevant to most cattle keepers are the legal provisions that refer to specific aspects of cattle husbandry on the farm, during transport, at market and at slaughter. These are diverse and numerous, but they are laid out within the DEFRA Code of Recommendations for the Welfare of Livestock: Cattle (2003), available free of charge.

Other important legislation that applies to cattle includes compliance with disease control regulations for notifiable diseases. These are diseases whose control is currently regulated by DEFRA in the UK. Those affecting cattle are:

- Anthrax.
- Bluetongue.
- Brucellosis.
- Bovine spongiform encephalitis (BSE).
- Contagious bovine pleuropneumonia.
- Enzootic bovine leukosis.
- Foot and Mouth disease.
- Lumpy skin disease.
- Rabies.
- Rift Valley fever.
- Rinderpest.

- Tuberculosis.
- Warble fly.

Regulations for control of cattle diseases include:

1. Registration of livestock holdings.

2. Cattle identification and registration with the British Cattle Movement Service.

3. Records of cattle movements, births and deaths.

4. Disease surveillance: DEFRA must immediately be notified if a notifiable disease is suspected. This includes sudden death (requiring anthrax investigation) and abortion (discretionary brucellosis sampling) in cattle.

5. Disease monitoring: all eligible cattle must be tested for TB and brucellosis as required by DEFRA. Brucellosis is also monitored via the milk from dairy herds. BSE samples are taken from cattle at abattoirs and (at the time of writing) through the Fallen Stock Scheme.

6. Mandatory cleansing and disinfection of livestock transport vehicles.

7. Ban on burial of livestock carcasses.

The public's safety from harm by cattle is protected by a range of legislation. For example, the Wildlife and Countryside Act 1981 bans bulls (uncastrated male cattle over ten months old) of recognized dairy breeds from being at large in fields crossed by rights of way. Other breeds of bull are also banned, unless with cows or heifers. The Animals Act 1971 provides for animals that are dangerous – for example cows protecting their calves or individuals that are aggressive – and makes the keeper of such an animal 'strictly liable' in most cases for injuries that the animal causes. Signs informing the public of the presence of cattle, particularly

*All cattle must be individually ear-tagged and have a passport.*

bulls, should be displayed. Health and Safety legislation requires that the public are not put at risk. The risks of working with cattle must be assessed and measures taken to comply with health and safety requirements.

Legislation designed to protect public health includes regulations on the use of medicines. A Code of Practice on the responsible use of animal medicines on the farm has been published by the Veterinary Medicines Directorate, providing guidelines about the use, storage and recording of farm animal medicines. It recognizes the need to plan ahead to prevent disease, to buy from authorized sources, to administer medicines properly, to observe withdrawal periods, to store medicines safely, to report side-effects, to dispose of unused medicines safely, and to record purchases and use of medicines (these records must be kept for five years). There are also waste management regulations that require all farm waste to be taken to a licensed site for disposal (unless exempted by licence issued by the Environment Agency).

# Glossary

**abomasum** The fourth stomach in ruminant animals. The abomasum functions like the stomach of non-ruminants.

**acidosis** Excess acid in the body, leading to disturbance of normal body function. Acidosis most commonly occurs in cattle as a result of excess acid production in the rumen with intake of too much highly fermentable feed such as barley or concentrate.

**afterbirth** *See* placenta.

**AI** *See* artificial insemination.

**artificial insemination (AI)** Delivery of semen into the uterus of the cow via a catheter. AI is used where it is impractical to keep a bull or to add new genetic material to the herd. Semen for AI in cattle is stored in 'straws' that are kept frozen in liquid nitrogen until required.

**anthelmintic** Medicine used in the control and treatment of parasitic worms that live in the gut or lungs.

**antibiotic** A medicine that inhibits growth of, or destroys, bacteria. Antibiotics are not effective against viruses.

**antibodies** Proteins produced by white cells in response to foreign molecules they meet. Each type of antibody is specific to a particular molecule, for example on the surface of harmful viruses or bacteria. Antibodies attach to the foreign material, preventing it from damaging or invading the host tissue. Antibodies are also produced in response to vaccination.

**bacteria** Single-celled, microscopic organisms that grow and multiply in living or inert material. There are many common diseases caused by bacteria.

**BCS** *See* body condition score.

**biosecurity** Protection of the herd from infectious and parasitic diseases that can be introduced via other livestock, wildlife, human contacts and vehicles or by air or water.

**body condition score (BCS)** Score (1–5 in the UK, 1–10 in the US) used to calibrate the level of fat cover in cattle.

**bolus** Lozenge, containing minerals or anthelmintic, introduced down the animal's throat to lodge in its rumen.

**bovine viral diarrhoea (BVD)** Viral infection that causes fertility problems and reduced immunity to other diseases such as pneumonia. Carrier animals (persistently infected or PIs), which are infected before birth and may be normal or chronically stunted, spread the disease.

**bulldog** Instrument for cattle restraint that fits across the skin between the nostrils. The animal keeps still so the bulldogs will not pinch its nose.

**BVD** *See* bovine viral diarrhoea

**calving interval (CI)** Number of days between one calving and the next. The target is generally 365 days.

**CI** *See* calving interval.

**cell count or somatic cell count (SCC)** Number of cells (mainly white blood cells derived from blood or lymph) per ml of milk. High cell counts (>200,000/ml) usually indicate udder infection, but cell counts also increase at the end of lactation.

**CL** *See* corpus luteum.

**corpus luteum (CL)** (Literally, yellow body.) Structure in the ovary that develops from the follicle once the ovum (egg) has been released. The CL produces the hormone progesterone.

**colostrum** Special udder secretion produced by the cow prior to, and for the first days after, calving. It is rich in antibodies, energy and protein. Udder secretion changes to milk within four days after calving.

**conception rate** Number of animals that conceive divided by the number that are served or inseminated.

**cudding (or ruminating)** The process whereby ruminants regurgitate feed from the rumen back to the mouth to chew it a second time and mix it further with saliva.

**culling** The removal of animals, generally breeding stock, from the herd, usually for slaughter at the end of their productive life. Culling may be voluntary, i.e. part of a planned programme to improve the efficiency or productivity of the herd, or forced, where cattle must be culled owing to disease or infertility.

**DCT** *See* dry cow therapy.

**DEFRA** Department for Environment, Food and Rural Affairs.

**DM** *See* dry matter.

**DMI** *See* dry matter intake.

**downer** Adult bovine animal that is unable to get up, usually as a result of injury (for example at calving) or metabolic disease.

**dry cow therapy (DCT)** Mastitis control treatment used when the cow is dried off at the end of lactation. The dry cow tube contains a long-acting antibiotic that is inserted into the teat to treat exisiting udder infection and to provide protection against new infection.

**dry matter (DM)** The dry matter content of a feed, measured as the total (fresh) weight less the weight of water it contains.

**dry matter intake (DMI)** The dry matter quantity of feed that an animal eats in a day.

**dystocia** Difficulty in calving.

**EBV** *See* Estimated Breeding Value.

**embryo** Developing young (calf) within the first few days after conception.

**Estimated Breeding Value (EBV)** Measure of the ability of a beef bull to confer improved genetic potential to offspring in terms of ease of calving and conformation.

**endometritis** Infection of the lining wall of the uterus.

**Farm Assurance Scheme (FAS)** A set of standards to which producers of meat, milk or eggs must comply. The standards generally include both food safety and animal welfare factors and have been developed by marketing bodies such as supermarkets. FAS may vary in the stringency of their demands and the rigour of their enforcement.

**FAS** *See* Farm Assurance Scheme.

**foetus** Developing young (calf) during pregnancy.

**follicle stimulating hormone (FSH)** The hormone that stimulates the development and ripening of a follicle, containing an egg, in the ovary.

**FSH** *See* follicle stimulating hormone.

**hormone** Chemical released into the blood that causes metabolic change(s) in one or more target organs. Examples include adrenaline released from the adrenal glands in times of acute stress to prepare the animal for 'fight or flight', and oestrogen, released from the developing follicle in the ovary to cause signs of bulling and to prepare the reproductive tract for service and conception.

**hybrid vigour** The phenomenon in which crossbred animals show higher genetic potential (e.g. for growth, disease resistance) than either of the purebred parents.

**IBR** *See* infectious bovine rhinotracheitis.

**immunity** Specific protection of the animal (or human) from disease. Humoral immunity is provided by antibodies circulating in the blood. Surface immunity is provided by antibodies on the membranes lining the gut, the respiratory and reproductive tracts. Cellular immunity is provided by particular white cells programmed to recognize and destroy foreign molecules.

**infectious bovine rhinotracheitis (IBR)** Viral disease causing fever, discharge from eyes and nose, coughing and reproductive disorders, mostly spread by carrier animals.

**killing out percentage** The proportion of a slaughtered animal that can be sold as meat, calculated as carcass weight divided by liveweight.

**LH** *See* luteinizing hormone.

**luteinizing hormone (LH)** Hormone that causes the egg to be released from the ripe follicle in the ovary, and the development of the corpus luteum.

**longevity** Length of life within the herd. Longevity is determined by the animal's health, fertility and productivity, and is a function of both genetic and management factors.

**liveweight gain (LWG)** Increase in body weight (usually measured as kg per day).

**LWG** *See* liveweight gain.

**mastitis** Inflammation of the udder tissue, usually caused by bacterial infection. Milk from affected animals is watery or clotted and its cell count increases. The udder may be hot and painful and the animal may be sick (toxic mastitis). Mastitis usually affects milking cows but may occur in dry cows ('summer mastitis'). Sub-clinically and chronically infected cows spread disease in the parlour via the milker's hands and milking machine. Cows may also be infected with mastitis organisms from the environment such as bedding.

**metabolism** The chemical processes that occur in living organisms. Examples include the build-up of products of digestion into new tissue for growth and the use of glucose and oxygen to provide energy for muscular work.

**metabolic disease** Imbalance of normal chemical processes, often occurring in early lactation in cattle. Examples include milk fever (imbalance of calcium metabolism), hypomagnesaemia (relative lack of magnesium) and ketosis/ acetonaemia (energy imbalance).

**metritis** Infection of the uterus that occurs soon (sometimes within twenty-four hours) after calving.

**microbe** Very small organism such as bacterium, yeast, virus or protozoon.

**mycoplasmas** Organisms similar to bacteria but lacking a rigid cell wall so they are susceptible to heat, disinfection and dessication. They can cause diseases such as pneumonia and are resistant to most antibiotics.

**oesophageal groove** A muscular channel in the wall of the rumen that closes when the calf suckles, effectively shutting off the first three stomachs so that milk is delivered directly into the fourth stomach, the 'true' stomach or abomasum (*see* diagram, page 52).

**oestrus** (Also called bulling or heat.) The period coinciding with release of the ovum (egg) when the heifer or cow is receptive to the bull and can become pregnant.

**ovulation** Release of the ovum (egg) from the ripe follicle in the ovary.

**ovum** Egg, produced in the ovary.

**oxytocin** Hormone produced in the brain that causes contraction of the uterus at calving and release of milk from the udder. Oxytocin acts remarkably rapidly, within two minutes of stimulation (e.g. of the calf trying to suckle).

**parity** Number of calves that a cow has produced.

**parturition** Calving.

**pathogen** Microbial organism able to invade, colonize and damage the tissues of plants or animals.

**pathogenicity** The ability of a microbe to invade, colonize and damage tissues of plants or animals.

**pH** A scale used to measure acidity, with 7 being neutral (neither acid nor alkali), numbers below 7 being progressively more acidic, and numbers above 7 being progressively more alkaline. Animals must keep their bodies at a constant pH for healthy cell function. Excess acid in the rumen (low pH) causes poor rumen function and metabolic disease.

**placenta** (Also called foetal membranes or afterbirth.) Fluid-filled sacs in which the foetal calf develops in the uterus. The placenta has areas of close contact between blood vessels of the mother and the foetal calf (cotyledons) to allow exchange of nutrients and oxygen from mother to calf, and waste products and $CO_2$ from foetus to mother. The placenta is expelled from the uterus once the calf has been delivered, in the third stage of calving.

**pregnancy rate** Number of animals that are pregnant divided by the number that were served or inseminated.

**protozoon** Single-celled organism, the simplest member of the animal kingdom. Certain cattle diseases are caused by protozoa, and protozoa also form a valuable part of the rumen microflora, breaking down plant material.

**predicted transmitting ability (PTA)** Measure of a dairy bull's ability to confer improved genetic merit (mainly milk production but also longevity and health) to offspring.

**PTA** *See* predicted transmitting ability.

**raddle** Harness with a coloured marker that fits under the chin or in front of the chest of the bull (or ram), so that any female that is served is marked and service date can be recorded.

**relative humidity (RH)** Level of moisture in the air expressed as a percentage of the maximum level of moisture it can carry at a particular temperature.

**RH** *See* relative humidity.

**rumen** The first of four stomachs in the ruminant animal. The rumen acts as a fermentation vat. The other stomachs are the reticulum, the omasum, and the abomasum.

**SCC** *See* cell count.

**scour** Diarrhoea.

**shear grab** Implement for precision-cutting and carrying forage from the silage face.

**slug feeding** Excessive and rapid intake of concentrate or TMR, likely to upset rumen fermentation and cause acidosis. Slug feeding occurs if high-yielding dairy cows or rapidly growing beef animals are suddenly offered the ration after a period without access to feed.

**slurry** Semi-fluid mix of urine and dung.

**steer/stirk** Castrated male bovine animal, usually reared for beef.

**store** Beef animal (or growing lamb) being fed to produce low growth rates during the winter.

**straights** Single feed components such as barley or soya, that are used in combination with forage to produce rations for livestock.

**sub-clinical disease** Disease or nutritional deficiency that does not cause clinical signs but may be associated with reduced production or resistance to disease, and may develop into the clinical condition. Examples include sub-clinical mastitis (associated with raised somatic cell count, reduced production and increased likelihood of clinical mastitis) and sub-clinical calcium deficiency (often followed by clinical milk fever or other diseases of early

lactation). Animals with sub-clinical disease may spread infection to others in the herd.

**total mixed ration (TMR)** A mix of forage(s) and straights, usually with added minerals and trace elements, that provides a complete ration formulated to meet the nutritional requirements of a group of cattle.

**TMR** *See* total mixed ration.

**trace element** Mineral that is essential for the animal's normal body function and health, required in very small quantities. Examples include copper, cobalt, iodine, selenium, zinc and manganese.

**transition cow** Dry (non-lactating) cow in the last two to three weeks prior to calving. Management of these animals is critical, especially in the high-yielding dairy herd, where cows experience very large metabolic changes as they calve and start lactation.

**urea** A compound produced from excess nitrogen derived from tissue breakdown or from the rumen. Urea is metabolized in the liver and excreted by the kidneys. Urea level in blood or milk can be measured to estimate nutritional status and/or kidney function.

**vaccination** Method for stimulating immunity against a particular disease. Vaccines may be live or killed, with live vaccines generally producing a more rapid effect than killed ones. Some last a lifetime, others must be boosted every one to two years. Vaccines are usually injected, but may also be given into the nose (e.g. against IBR), by mouth (e.g. against lungworm), or by scraping the skin (e.g. orf vaccine in sheep).

**virus** Minute organism, much smaller than bacteria, that can that multiply only in living plant or animal cells. Many diseases are caused by viruses.

**white cells** Specialized cells whose job is to protect the animal or human from disease. Some produce antibodies; others circulate in the bloodstream and through body tissues to neutralize foreign material such as bacteria.

**withdrawal period** The time after administration of any drug to a food-producing animal, during which its milk or meat cannot be sold for human consumption.

# References and Further Reading

## CHAPTER 1

### Dairy Farming Systems

Brand, A., Noorhuizen, J. P. T. M., Schukken, Y. H. (2001) *Herd Health and Production Management in Dairy Practice* (Wageningen Press).

MDC (2002) 'Longevity – controlling culling to improve herd profitability.' *Research into Practice*, publication number 51.

Sibley, R. (2006) 'Developing health plans for the dairy herd.' *In Practice* 28; pp.114–21.

## CHAPTER 2

### Beef Farming Systems

Allen, D. (1990) *Planned Beef Production and Marketing* (Blackwell Science Ltd).

Bebbington, T. (2006) 'Health planning for beef suckler herds.' *In Practice* 28; pp.370–5.

Brown C., Heal J., and Powdrill S. (2005) 'Targeted beef selection and handling.' English Beef and Lamb Executive (EBLEX).

Cutler K. (2003) 'The trace element nutrition of beef suckler cattle.' *Cattle Practice:* 11

DEFRA (2002) 'Condition scoring of beef suckler cows and heifers.'

Evans D. (2005) *Future Trends in the European Beef Industry: A Global View* (National Beef Association).

Food Standards Agency, ADAS (2004) *Clean Beef Cattle for Slaughter: A Guide for Producers.*

Pullar, D. (2002) 'Making the market work for you.' *MLC* article 2.

Topliff, M. (2007) 'The outlook for the UK beef sector.' (MLC outlook conference proceedings).

## CHAPTER 3

### The Calf

Bazeley K. (2003) 'Investigation of diarrhoea in the neonatal calf.' *In Practice* 25; pp.152–159

De Passille, Rushen and Weary (2004) 'Designing good environments and management for calves.' *Advances in Dairy Technology* 16; pp.75–89

Drackley, J. K. (2005) 'Early growth effects on subsequent health and performance of dairy heifers.' In Garnsworthy, P.C. (ed.) *Calf and Heifer Rearing*; pp.213–36 (Nottingham University Press).

Grove-White, D. H. (1998) 'The first two weeks of life – a high risk period.' *Cattle Practice* 6.

Quigley J., et al. (2006), (2005) 'Passive immunity in newborn calves.' In Garnsworthy, P.C. (ed.) *Calf and Heifer Rearing*; pp.135–58 (Nottingham University Press).

Thickett, B., Mitchell, D., and Hallows, B. (1986) *Calf Rearing* (Crowood Press).

Straiton E. (2002) *Calving the Cow and Care of the Calf* (Crowood Press).

## CHAPTER 4

### Heifer Rearing

Allen, D., Kilkenny, B. (1984) *Planned Beef Production* (2nd edn) ISBN 0246121947 (Collins).

Heinrichs, A. J., and Hutchinson, L. J. 'Management of dairy heifers.' *Penn State Extension Circular 385* (The Pennsylvania State University. www.das.psu.edu/dairynutrition/heifers).

## CHAPTER 5

### Nutrition

Albright, J. L. (1993) 'Feeding behaviour of dairy cattle.' *Journal of Dairy Science* 76; pp.485–98.

Albright, J. L., and Arave, C. W. (1997) 'Feeding behaviour.' In *The Behaviour of Cattle;* p. 106 (CAB International, Wallingford, Oxon, UK).

Allen, D., Kilkenny, B. (1984) *Planned Beef Production* (2nd edn) (Collins).

Bickert, W. G. (1990) 'Feed Manger & Barrier Design.' *Dairy Feeding Systems* NRAES–38; pp.199–206 (Northeast Regional Agricultural Engineering Service).

Cook, N. B., Nordlund, K.V., and Oetzel, G.R. (2004) 'Solving Fresh Cow Problems: the Importance of Cow Behavior.' In *Proceedings of the 8th Dairy Symposium of the Ontario Large Herd Operators;* pp.248–55.

Brouk, M. J., Smith, J. F., and Harner, J. P. (2003) 'Effect of feedline barrier on feed intake and milk production of dairy cattle.' In Janni, K. A. (ed.) *Proceedings of the Fifth International Dairy Housing Conference;* pp.192–5 (American Society of Agricultural and Biological Engineers, St Joseph, Michigan).

Chamberlain, A. T. and Wilkinson, J. M. (1996) *Feeding the Dairy Cow* (Chalcombe Publications).

Dado, R. G., and Allen, M. S. (1994) 'Variation in and relationships among feeding, chewing and drinking variables for lactating dairy cows.' *Journal of Dairy Science* 77; pp.132–44.

DeVries, T. J., von Keyserlingk, M. A. G., and Weary, D. M. (2004) 'Effect of feeding space on the inter-cow distance, aggression, and feeding behaviour of free-stall housed lactating dairy cows.' *Journal of Dairy Science* 87(5); pp.1432–8

Friend, T. H., and Polan, C. E. (1974) 'Social rank, feeding behaviour and free-stall utilization by dairy cattle.' *Journal of Dairy Science* 57; pp.1214–22.

Friend, T. H., Polan, C. E., and McGiliard, M. L. (1977) 'Free-stall and feed bunk requirements relative to behaviour, production and individual feed intake in dairy cows.' *Journal of Dairy Science* 60; pp.108–18.

Grant, R. J., and Albright, J. L. (1995) 'Feeding behaviour and management factors during the transition period in dairy cattle.' *Journal of Animal Science* 73; pp.2791–803.

Grant, R. J., and Albright, J. L. (1997) 'Dry matter intake as affected by cow grouping and behaviour.' In *Proceedings of the 58th Minnesota Nutrition Conference;* pp.93–103 (University of Minnesota, Bloomington, MN).

Grant, R. J., and Albright, J. L. (2000) 'Feeding Behaviour.' In D'Mello, J. P. F. (ed.), *Farm Animal Metabolism and Nutrition* (CAB International, Wallingford, Oxon, UK).

Grant, R. J., and Albright, J. L. (2001) 'Effect of animal grouping on feeding behaviour and intake of dairy cattle.' *Journal of Dairy Science* 84 (E. Suppl.); E156–E163.

Grummer R. R., Mashek, D. G., Hayirli, A. (2004) 'Dry matter intake and energy balance in the transition period.' *Veterinary Clinics of North America Food and Animal Practice*, Nov 20 (3); pp.447–70.

Hasegawa N., Nishiwaki, A., Sugawara, K., Ito, I. (1997) 'The effects of social exchange between two groups of lactating primiparous heifers on milk production, dominance order, behaviour and adrenocortical response.' *Applied Animal Behavioural Science* 51: pp.15–27.

Ingvartsen, K. L., Friggens, N. C., and Faverdin, P. (1999) 'Food Intake in late preganancy and early lactation.' In *Metabolic Stress in Dairy Cows*, Occasional Publication No. 24. (British Society of Animal Science)

von Keyserlingk, Marina and DeVries, Trevor (2004) 'Designing better environments for cows to feed.' *Advances in Dairy Technology* 16; pp.65–73.

Kondo, S., Sekine, J., Okub,o M., and Asahida, Y. (1989) 'The effect of group size and space allowance on the agonistic spacing behaviour of cattle.' *Applied Animal Behavioural Science* 24; pp.127–35.

Kondo, S and Hurnik J. F. (1990) 'Stabilization of social hierarchy in dairy cows.' *Applied Animal Behavioural Science* 27; pp.287–97.

Manson, F. J., and Appleby, M. C. (1990) 'Spacing of dairy cows at a feed trough.' *Applied Animal Behavioural Science* 26; pp.69–74.

Menzi, W., Jr., and Chase, L. E. (1994) 'Feeding behaviour of cows housed in free-stall barns.' In *Dairy Systems for the 21st Century*; pp.829–31 (American Society of Agricultural and Biological Engineers, St Joseph, Michigan).

Mertens, D. R. (1994) 'Regulation of forage intake.' In Fahey, G. C. (ed.), *Forage Quality, Evaluation, and Utilization*; pp. 450–93 (American Society of Agronomy, Madison, Wisconsin).

Murphy, M. R., Davis, C. L., and McCoy, G. C. (1983) 'Factors affecting water consumption by Holstein cows in early lactation.' *Journal of Dairy Science* 66; pp.35–8.

Sanders, D. E. (1990) *Boosting Dairy Profits* (American Veterinary Publications, Goleta, California).

Smith, J. F., Harner III, J. P., Brouk, M. J., Armstrong, D. V., Gamroth, M. J., and Meyer, M. J. (2000) 'Relocation and expansion planning for dairy producers.' *Publ. MF2424.* (Kansas State Univ. Coop. Ext. Serv., Manhattan).

Sniffen, C. J. (1991) 'Grouping management and physical facilities.' *Veterinary Clinics of North America Small Animal Practice* 7; pp.465–78.

Soriano, F. D. (1998) 'Grazing and feeding management of lactating dairy cows.' *M.S. Thesis* (Virginia Polytechic Institute, Blacksburg).

# CHAPTER 6

## Grass and Grazing Management

Depies, K. K. (1994) 'The effect of intensive rotational stocking on the nutrient utilization of lactating dairy cows.' *M.S. Thesis* (University of Wisconsin, Madison).

DEFRA and RPA (2006) 'Single farm payment scheme cross compliance guidance for soil management.'

Holmes, W. (1980) 'Grazing management.' In Holmes, W. (ed.), *Grass: Its Production and Utilization*; pp.125–73 (Blackwell Scientific Publications, Boston, Mississippi)

Holmes, W. (1989) 'Application on the farm.' In Holmes, W. (ed.), *Grass: Its Production and Utilization* (2nd edn); pp.258–71 (Blackwell Scientific Publications, Boston, Mississippi)

Mayne, C. S., Newberry, R. D., Woodcock, S. C. F., and Wilkins, R. J. (1987) 'Effect of grazing severity on grass utilization and milk production of rotationally grazed dairy cows.' *Journal of British Grassland Society* 42; pp.59–72.

Harland J. I., *MDC Grass+ Grassland Management Improvement Scheme* (CD-ROM.) Contributors: C. Fox, A. Bailey, J. Bax, S. Brandon, I. Browne, J. Harland, G. Lane, S. Mayne, R. Simpson (Milk Development Council).

Muller, L. D. 'Managing to Get More Milk and Profit from Pasture.' www.das.edu/dairynutrition/forages/pasture

Muller, L. D., Delahoy, J., and Bargo, F. 'Supplementation of lactating cows at pasture.' www.das.edu/dairynutrition/forages/pasture

Muller, L. D., Sullivan, K., and Sodera, K. 'Supplementing pasture with a total mixed ration.' www.das.edu/dairynutrition/forages/pasture

Voisin, A. (1988) 'Law of rotational grazing.' In Voisin, A. (ed.) *Grass Productivity*; pp.131–4 (Island Press, Washington, DC).

# CHAPTER 7

## Management of Fertility

ADAS/IGR University of Bristol (2001) 'Beef Breeding.' Livestock Knowledge Transfer (factsheet).

Andrews, A. H. (ed.), with Blowey, R. W., Boyd, H. and Eddy, R. G. (2004) *Bovine Medicine: Diseases and Husbandry of Cattle* (2nd edn) (Blackwell Science Ltd).

British Limousin Cattle Society 'Performance Recording: Understanding Estimated Breeding Values.' (www.limousin.co.uk/breed/sire_dam_summary).

Caldow, G., Lowman, B., and Riddell, I. (2005) 'Veterinary intervention in the reproductive management of beef cow herds.' *In Practice* 27.

Esslemont, R. J. (2003) 'The costs of poor fertility and what to do about reducing them.' *Cattle Practice* 11.

Logue, D. (2005) 'Problems with the bull.' *Cattle Practice* 13.

Lowman, B. (2004) 'Estimated breeding values for beef cattle.' *In Practice* 26; pp.206–11.

McGowan, M. (2004) 'Approach to conducting bull breeding soundness examinations.' *In Practice* 26; pp.485–91.

PD Advisory Panel: R. Holland, H. Black, R. Esslemont, P. Latham, R. Webster, M. Glover, R. Sibley, M. Warren (2003) *MDC PD+ Fertility Improvement Programme* (MDC).

# CHAPTER 8

## Housing

BS 5502: Part 40 (1990) 'Code of practice for design and construction of cattle buildings.' *Buildings and Structures for Agriculture* (British Standards Institute).

Blowey R. and Edmondson P. (2000) 'The Environment and Mastitis.' *In Practice* 22; pp.382–95.

Bohmanova J., Misztal, I., and Cole, J. (2007) 'Temperature-humidity indices as indicators of milk production losses due to heat stress.' *Journal Of Dairy Science* 90; pp.1947–56.

J DEFRA, PB 7949 (2003) 'Code of recommendations for the welfare of livestock (cattle).'

DEFRA/ADAS (2006) 'Action on animal health and welfare. Housing the modern dairy cow.'

Grandin, T. (2007) *Livestock Handling and Transport* (CAB International, Wallingford, Oxon, UK).

Hartung, J. (1994) 'Environment and animal health.' In Wathes, C. M. and Charles, D.R. (eds) *Livestock Housing* (CAB International, Wallingford, Oxon, UK).

Hicks B. (2006) 'Effect of management strategies on reducing heat stress in feedlot cattle.' (Oklahoma Cooperative Extension Service Beef Cattle Research Update.)

Hughes, J. (2000) 'Cows and Cubicles.' *In Practice* 22; pp.230–9

Kelly, M. (2005) 'General checklist for dairy unit design and construction.' *UK Vet* 10; pp.55–9.

Kelly, M. (2004) 'A practical approach to good cattle housing.' *UK Vet* 9; pp.36–40

Kelly, M. 'Calf housing ventilation – upgrading an existing building.' *UK Vet* 8; pp.26–9.

Kelly, M. 'Good ventilation is essential for livestock housing.' *UK Vet* 8; pp.33–6.

Kelly, M. (2006) 'Minimizing slurry pooling in dairy housing.' (MDC).

Langley R.W. (2006) *Making the Most of Your Farm Buildings* (Crowood Press).

MDC (2006) *Housing the 21st-Century Cow* (DVD).

Mee, A. (1990) 'Dystocia in Friesian Heifers.' *Veterinary Record* 127; pp.219–24.

Miller, K., and Wood-Gush, D. G. M. (1991) 'Some effects of housing on the social behaviour of dairy cows.' *Animal Production* 53; pp.271–8.

Moran, J. (2005) 'Temperature–Humidity Index.' In *Feeding Management For Smallholder Dairy Farmers in the Humid Tropics* (Landlinks Press).

Offer, J., Kelly, M., Logue, D., Mason, C., Barrett, C., Howat, D. (2006) 'Effective footbathing of dairy cows.' *Research into Practice* (MDC).

Webster, A. J. F. (1984) *Calf Husbandry, Health and Welfare* (Granada, London).

Webster, A. J. F. (1994) 'Comfort and injury.' In Wathes, C. M. and Charles, D. R. (eds), *Livestock Housing*; pp.49–68 (CAB International, Wallingford, Oxon, UK).

## CHAPTER 9

### Health and Disease

Blowey, R. W. (1985) *The Veterinary Book for Dairy Farmers* (Farming Press).

Blowey, R. W. and Edmondson, P. (1995) *Mastitis Control in Dairy Herds: Illustrated and Practical Guide* (Old Pond Publishing).

Huxley, J. (2006) 'Assessment and management of the recumbent cow.' *In Practice* 28; pp.176–84

MDC and British Cattle Veterinary Association (2006) 'Cattle purchasing checklist.' *Research into Practice.*

MDC (2006) 'Mastitis cell counts.' *Research into Practice.*

Rebhun, W. C. (1995) *Diseases of Dairy Cattle*; pp.1–12 (Williams and Wilkins, USA).

Straiton, E. (2000) *Cattle Ailments* (Crowood Press).

Touissant Raven, E. (1989) *Cattle Footcare and Claw Trimming* (Crowood Press).

Watson C. (2006) *Lameness in Cattle* (Crowood Press).

Whay, H. R. (1999) 'The way cattle walk: steps towards lameness control.' *Cattle Practice* 7.

## CHAPTER 10

### Welfare and the Law

Dawkins, M. S. (1998) 'Animal suffering – the biobehavioural background.' In Mitchell, A. R. and Ewbank, R. (eds), *Ethics, Welfare Law and the Market Forces: the Veterinary Interface*; pp.1–12 (RCVS and UFAW).

DEFRA, PB 7949 (2003) *Code of Recommendations for the Welfare of Livestock (Cattle).*

FAWC (1997) *Report on the Welfare of Dairy Cattle.*

Health and Safety Executive (2006) *Cattle and Public Access in England and Wales.*

Hemsworth, P. H. and Coleman, G. J. (1998) *Human–Livestock Interaction: Stockpersons and the Productivity and Welfare of Intensively Farmed Animals*, ISBN 0851991955 (CAB International, Wallingford, Oxon).

Main, D. and Cartledge, V. (2000) 'Farm assurance schemes: what is the veterinarian's role?' *In Practice* 22; pp.335–9.

McInerney, J. P. (1998) 'The economics of welfare.' In Mitchell, A. R. and Ewbank, R. (eds), *Ethics, Welfare Law and the Market Forces: the Veterinary Interface*; pp.115–34 (RCVS and UFAW).

Radford, M. (2001) *Animal Welfare Law in Britain* ISBN 0-19-826245-40 (Oxford University Press).

Seabrook, M. (2000) Proceeding of 3rd NAHWOA Workshop *Human–Animal Relationships: Management, Housing and Ethics.*

Sibley, R. (2006) 'Developing health plans for the dairy herd.' *In Practice* 28; pp.114–21.

Webster, John (1996) *UFAW Animal Welfare: A Cool Eye Towards Eden* Blackwell Science, UK).

Webster, John (2006) *UFAW Animal Welfare: Limping Towards Eden* (Blackwell Science, UK).

# Index